# FAULT LOCATION IN ELECTRONIC EQUIPMENT: AN INTRODUCTION

K. J. Bohlman
I.Eng., F.I.E.I.E., A.M.Inst.E.

Dickson Price Publishers Ltd
Hawthorn House
Bowdell Lane
Brookland
Romney Marsh
Kent TN29 9RW

Dickson Price Publishers Ltd
Hawthorn House
Bowdell Lane
Brookland
Romney Marsh
Kent TN29 9RW

First published 1991
© K. J. Bohlman 1991

British Library Cataloguing in Publication Data
Bohlman, K. J. (Kenneth John)
   Fault location in electronic equipment
   1. Electronic equipment. Faults. Detection
   I. Title
   621.381

   ISBN 0-85380-132-0

*All rights reserved. No part of this publication may be reproduced, stored in a retrieval system, or transmitted in any form or by any means electronic, mechanical, photocopy, recording or otherwise, without the prior permission of the publishers.*

Photoset by
R. H. Services, Welwyn, Hertfordshire
Printed and bound in Great Britain by
Billing & Sons Ltd, Worcester

# CONTENTS

|   |                                 |    |
|---|---------------------------------|----|
|   | Preface                         | v  |
| 1 | Fault Location Methods          | 1  |
| 2 | Scheme for Fault Location       | 11 |
| 3 | Static Tests                    | 15 |
| 4 | Fault Tracing in Logic Systems  | 46 |
| 5 | Fault Location Exercises        | 55 |

## Other Books of Interest

Electronics Servicing Vol 1
Electronics Servicing Vol 2
Electronics Servicing Vol 3
Electronics Servicing 500 Q & A for Part 1
Electronics Servicing 500 Q & A for Part 2
Control Systems Technology
Digital Techniques
Principles of Domestic Video Recording and Playback Systems
Video Recording and Playback Systems 500 Q & A
Colour and Mono Television Vol 1
Colour and Mono Television Vol 2
Colour and Mono Television Vol 3
Radio Servicing Vol 1
Radio Servicing Vol 2
Radio Servicing Vol 3

### Inspection Copies

Lecturers wishing to examine any of these books should write to the publishers requesting an inspection copy.

Complete Catalogue available on request.

# PREFACE

THIS BOOK COMMENCES with an introduction to fault diagnosis methods and dynamic testing and outlines a proven scheme for fault location which may be applied to a wide range of electronic equipment. In-circuit and out-of-circuit testing of active/passive components and static voltage and resistance measurements are described. Both analogue and digital fault-finding techniques are discussed, with the main emphasis on the application to simple circuits and systems. Fault location exercises with answers are also included.

Students, particularly those following Parts 1 and 2 of the City and Guilds 224 course or professionals and amateurs seeking introductory guidance in general fault-finding techniques will find the information in this book useful.

CHAPTER ONE

# FAULT LOCATION METHODS

IN ADDITION TO an understanding of circuit and equipment operation, engineers and technicians engaged in the servicing of industrial or domestic electronic equipment also need to develop some form of basic logical approach to fault diagnosis that can be applied to a wide range of electronic apparatus and systems. A number of different methods for fault diagnosis exist but only those which adopt a systematic approach, see Fig. 1.1, will be considered in this book. A **systematic** method is one which is governed by a set of rules following a logical approach to fault detection with the rules applied sequentially, i.e. one after the other without missing steps.

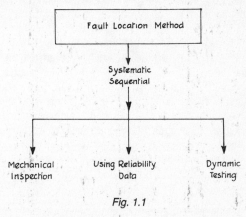

Fig. 1.1

## USING RELIABILITY DATA

The reliability of components within an electronic unit or units within a system are often known by equipment and component manufacturers; the data being compiled from in-house testing and field failure reports. When a fault occurs, a decision to change a particular component or unit may be made on the basis of available reliability data.

# 2 FAULT LOCATION IN ELECTRONIC EQUIPMENT

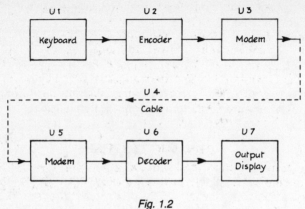

*Fig. 1.2*

Consider the individual units of the sequential digital data system given in Fig. 1.2. Reliability data may be available which shows that of the last 100 unit failures, 52 have occurred in U3, 33 in U5, 12 in U7, 3 in U1 and none in the remaining units. Thus on the basis of the reliability data, units U3, U5, U7 and U1 may be changed first in that order in an attempt to clear the fault. Of course, changing these particular units may fail to clear the fault and further testing or replacement will need to be considered.

The same principle may be applied to circuit components within a unit and on the basis of reliability data, components which have been shown to fail more often than others may be replaced first in an attempt to clear the fault.

Fault location based on reliability data is well established in many industries where repair technicians use manufacturers and personal reliability data to build up a knowledge of 'stock faults' and their causes. Apart from fault detection, reliability data may be used to improve the 'availability' of a system by duplicating vulnerable component units.

The strengths of this method of fault location include the speed at which faults may be corrected (vital services need to be restored quickly), low labour costs and the use of semi-skilled personnel. Its main weakness is that, although systematic in approach, it is limited by the amount and detail of the reliability data available and there is no plan for 'what to do next' when suggested unit or component replacements are exhausted.

## MECHANICAL INSPECTION

This is purely a visual method of fault detection and is useful if carried out systematically. It is well known that failures on a component printed circuit board may be due to hair-line cracks in the printed circuit, dry soldered connections or short-circuited conductors. Thus a visual inspection of an illuminated p.c.b. from the print side with the aid of a magnifying lens and working across the board from one end to the other may often assist in pin-pointing faults of this nature. Similarly from the component side of the

# FAULT LOCATION METHODS

p.c.b., visual inspection may be carried out systematically looking for damage to components.

This method can be very quick and useful especially under catastrophic failure conditions. It should be noted that *a burnt-out component is not necessarily the cause of the fault condition and the fault may lie elsewhere* in the circuit; for example, a blown fuse seldom indicates a faulty fuse. In these circumstances it is essential to refer to a circuit diagram and perhaps perform additional tests/measurements to ascertain the likely cause.

## DYNAMIC TESTING

'Dynamic testing' is concerned with the application of a logical sequence of tests to the individual blocks or stages of a system with the equipment in the 'on' mode, following the 'signal' route through from beginning to end. The technique is most suited to series-connected units or stages to which many electronic systems conform in whole or part. It can be applied to analogue, pulse or digital series-connected systems and is arguably the most powerful fault location method for general use.

Consider the general series-connected system shown in Fig. 1.3 consisting of six stages or units A–F lying in a 'signal' path. If it is assumed that a single fault is present in one of the stages, one method of locating the faulty stage or unit is to connect a monitor (a c.r.o., voltmeter or probe, etc.) at the output of Block F and to inject a suitable test signal into each of the test points in turn, commencing at TP6 and working back to TP1 while assessing the monitor for the correct response each time the signal is injected.

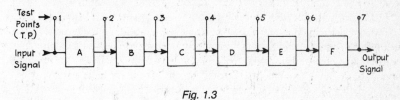

Fig. 1.3

Alternatively, a suitable test signal may be applied at TP1 and the monitor moved progressively from TP2 to TP7 to check for the correct output response from each stage or unit.

In principle, the two methods are similar in that dynamic testing eliminates each stage in turn. The particular method chosen will depend upon the type of electronic system, the suitability of available signal injection and monitoring equipment and any loading effects encountered.

### Dynamic Testing Examples
(1) *Audio Amplifier*

Figure 1.4 shows a common method for the dynamic testing of an audio amplifier which normally receives an input from a tape player, record player

# 4 FAULT LOCATION IN ELECTRONIC EQUIPMENT

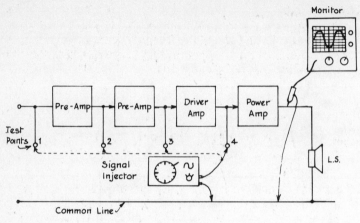

Fig. 1.4

or microphone to produce sound in the loudspeaker. A c.r.o. is used as a 'signal monitor' and a standard signal generator as the 'test signal source'. An a.c. voltmeter could alternatively be used as the monitor but a c.r.o. is better as it will show up any tendency to 'overdrive' when injecting the test signal and will highlight the early onset of stage distortion.

To locate the **faulty stage**, a test signal at, say, 1kHz is injected into test point 4 at an amplitude that would normally produce a standard output power of, say, 100mW. If the nominal impedance of the loudspeaker is 10Ω, the r.m.s. voltage across the loudspeaker would be given by

$$\sqrt{(R.P)} = \sqrt{(10 \times 100 \times 10^{-3})} = 1V \text{ r.m.s.} = 1.414V \text{ peak.}$$

Thus with 100mW of signal power dissipated in the loudspeaker, the display on the c.r.o. would be 2.824V peak-to-peak (sinewave signal). Therefore with the **correct level** of signal injected into test point 4, a sinewave signal of approximately 2.8V peak-to-peak will be displayed provided the stage is functioning normally. If no output, low output or distorted output is obtained then the fault probably lies in the power amplifier.

By moving the test signal source from TP4 towards TP1 while progressively **reducing** the injected signal level and observing the c.r.o. display, the particular stage that is faulty may be located.

The successful application of this dynamic testing method requires a knowledge of the expected signal injection level or 'sensitivity' at each stage input for the particular amplifier under test in order to raise the chosen standard output power. Some idea of a typical 'average value' for the signal injection level may be obtained by tabulating the results for each test point on a number of similar types of serviceable amplifiers. If during fault finding, an excessively large input is required at a particular test point to raise the standard output it may be concluded that a fault exists between that test point and the last 'good' test point.

(2) *Pulse Circuit*

Dynamic testing of a pulse circuit is illustrated in Fig. 1.5 which shows a simple series-connected pulse system. Unlike the audio amplifier, the first stage does not require an input to make the system function; the first stage (the crystal oscillator) is the signal source.

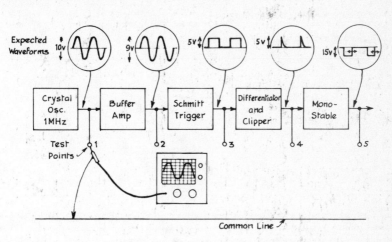

Fig. 1.5

When the circuit under test is of this nature, a signal injection source is thus not normally required. Since the correct functioning of each stage is indicated by its expected waveform output, it is more convenient to **move the monitor** (a c.r.o.), commencing at TP1 and moving progressively towards TP5. At each test point the displayed waveform is compared with the expected waveform and in this way the stage or area that is faulty may be located. For example, if the waveform at TP2 is correct but there is no output at TP3, then the fault probably lies in the Schmitt trigger stage.

Care should be taken when assessing waveforms for correctness and, in general, we should considered features such as **Shape–Amplitude–Frequency–Pulse width–D.C. component**, etc.

(3) *Radio Receiver*

A method similar to that used for the audio amplifier may be applied to the dynamic testing of a radio receiver, see Fig. 1.6.

A c.r.o. may be connected across the loudspeaker to act as the monitor and an appropriate frequency test signal applied to the test points commencing at TP6 and working back to TP1, with the level of the injected signal adjusted to raise a standard output power. The frequency of the test signal and typical injected signal levels required to raise a standard output of 50mW are given in Table 1.1.

# 6 FAULT LOCATION IN ELECTRONIC EQUIPMENT

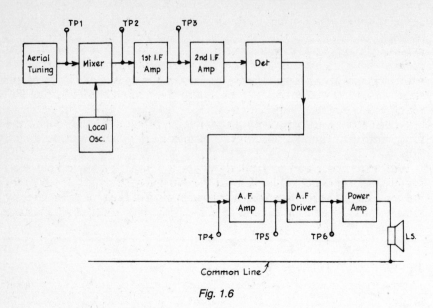

Fig. 1.6

Table 1.1 Test Signal Frequencies for Dynamic Testing of A.M. Transistor Radio

| Frequency of injected signal | Level (sensitivity) | Test point |
|---|---|---|
| 400Hz | 600mV | TP6 |
| 400Hz | 50mV | TP5 |
| 400Hz | 5mV | TP4 |
| Mod. i.f. (470kHz) | 3mV | TP3 |
| Mod. i.f. (470kHz) | 200µV | TP2 |
| Mod. r.f. (200kHz, LW) | 30µV | TP1 |
| Mod. r.f. (1MHz, MW) | 30µV | TP1 |

For the testing of the audio stages of the receiver, a test tone lying in the range 400Hz–1kHz is commonly used with the injected signal level progressively reduced in moving from TP6 towards TP4.

The i.f. stages may be tested dynamically using a modulated i.f. signal with the frequency set to suit the i.f. of the receiver in question (typically lying in the range 455–475kHz).

To test the operation of the 'frequency changer', modulated test signals at two or three spot frequencies on both the LW and MW bands should be used so that the sensitivity across each band may be checked.

If normal sensitivity is obtained at TP2 but no output or low output is

# FAULT LOCATION METHODS

experienced when injecting at TP1 on MW and LW, then the fault lies either in the Mixer or Local Oscillator stages. The Local Oscillator may be tested dynamically by connecting the c.r.o. to its output and checking for the presence of a continuous sinewave oscillation of appropriate frequency which should change frequency as the receiver tuning is altered. The Mixer stage may be tested independently by injecting modulated i.f. into TP1 which will check its ability to 'amplify'.

When carrying out sensitivity checks it is important to set the volume and tone controls to known positions, e.g. maximum volume and maximum 'top'.

With experience and particularly under no output or low output fault symptoms it is possible to implement the dynamic tests by just listening for the test tone in the loudspeaker; a change in pitch of the tone when altering the injection point may indicate excessive injection signal level or stage distortion.

### The Half-split Technique

Although methodical, dynamic testing of a system stage-by-stage is a slow process, particularly in a series chain consisting of a large number of stages or units. The **half-split** technique reduces the number of tests that need to be made to isolate the faulty stage or unit by making each test at such a point in the system **that half the number of units is eliminated at each test**.

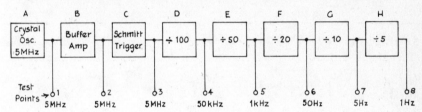

*Fig. 1.7*

Consider the application of the half-split technique to the frequency divider chain of Fig. 1.7 where a single fault is known to exist in one of the stages. The sequence of tests would be:

(a) Split the system into half by measuring the output at TP4 using, say, a frequency counter or c.r.o. If output is obtained, the fault lies in Blocks E–H; if not, the fault lies in Blocks A–D. Suppose that output is obtained at TP4.
(b) Split Blocks E–H into half by measuring the output at TP6. If output is obtained, the fault lies somewhere in Blocks G–H; if not, the fault is confined to Blocks E–F. Suppose that output is not obtained at TP6.
(c) Split Blocks E–F into half by measuring the output at TP5. If output is obtained, the fault lies in Block F; if not, the fault lies in Block E. Suppose that output is obtained. Thus the fault lies in Block F.

# 8  FAULT LOCATION IN ELECTRONIC EQUIPMENT

It will be seen that only three tests are required to isolate the faulty block, as opposed to a maximum of eight tests if block-by-block tests are made. For a series chain consisting of 100 blocks only seven tests would be required, which illustrates the superiority of the half-split technique over block-by-block testing.

It may not always be possible to split a system into an equal number of blocks, in which case the nearest to the half-split should be used. Also, particularly in complex series systems, it may be more difficult to make tests at certain points than at others because of loading effects, the tests may be very time consuming or specialist test equipment is required. Only a single fault was assumed to have existed in Fig. 1.7, but even when multiple faults are present the half-split technique is still the most efficient method to use.

If the half-split technique is applied to the radio receiver of Fig. 1.6, the system could be roughly split in half by making the first test at TP4. If output is obtained in the loudspeaker when signal is injected into TP4 (at the correct level), the fault is confined to earlier stages; if not, the fault lies in the audio stages of the receiver.

## Branching and Feedback in a Series System

Electronic systems consisting of series-connected stages or units often contain parallel branches and sometimes feedback loops which complicate fault location.

### *Divergence*

Branching is said to be 'divergent' when an output from one stage or unit feeds two or more other stages or units, see Fig. 1.8, where it is assumed that the input to block A is sustained from earlier blocks.

By inspecting the 'signal' outputs at TPX and TPY, the following deductions can be made:

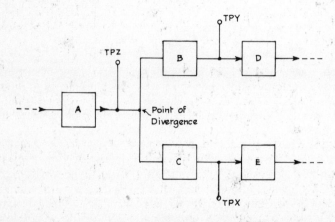

*Fig. 1.8*

(a) Outputs at TPX and TPY both correct – fault lies in one of the following series paths in which Block C or Block E is included.
(b) Output at TPX correct but output at TPY incorrect – fault lies probably in Block B.
(c) Output at TPY correct but output at TPX incorrect – fault lies probably in Block D.
(d) Outputs at TPX and TPY both incorrect – fault lies probably in Block A or earlier stage in which case the output at TPZ will be incorrect.

*Convergence*

Branching is said to be 'convergent' when two or more input lines are fed to a stage or unit, see Fig. 1.9. Here the output from Block B will be correct only when all three inputs f, g and h are correct.

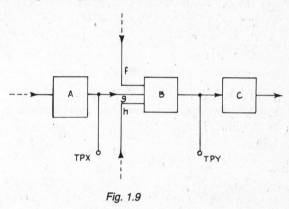

Fig. 1.9

If it is found that the output at TPX is correct but no output or incorrect output is obtained at TPY, **all three inputs** f, g and h must be tested individually. If, for example, input f is found to be incorrect or non-existent, the fault lies in the circuit producing input f or its feed line to Block B. Should all three inputs be correct, the fault lies in Block B.

*Feedback*

Electronic systems with feedback loops which connect the output of a particular block to an earlier block in the system via some form of 'signal' network can, with certain faults, present one of the more difficult problems to fault location.

Often the output of the final block is fed to an earlier block causing a closed loop to exist as in Fig. 1.10. In some cases, the effect of the feedback is only to **modify** the system's performance, but in other cases output will be obtained only if the feedback is present, i.e. its effect is to **sustain** an output of some kind of oscillation or a steady level. A.C. negative feedback used in an amplifier to reduce distortion, increase bandwidth or modify input/output impedance or feedback along the a.g.c. line of a radio receiver are

# 10 FAULT LOCATION IN ELECTRONIC EQUIPMENT

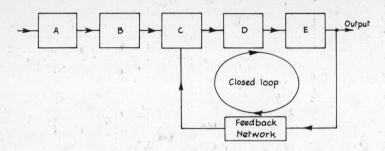

*Fig. 1.10*

examples of 'modifying' feedback. The positive feedback in an oscillator or the feedback in a position control servo system are examples of 'sustaining'-type feedback.

When the feedback is of the 'sustaining' type and a single fault is present within the closed loop, all of the stages within the loop may appear to be faulty, for example, d.c. voltages on all stages show up as incorrect. This problem arises in d.c.-coupled stages and is compounded when d.c. feedback is used to stabilise operating conditions. Since the fault may lie within the feedback network or an incorrect signal is fed back from the faulty stage, it is advantageous to disconnect the feedback path and substitute a suitable feedback signal so that testing of the individual stages may be carried out. To perform this safely requires a good knowledge of the operation of the system under test.

With 'modifying' feedback systems, open-circuiting of feedback lines is sometimes more feasible but care must be taken to avoid upsetting d.c. conditions. In some cases it may be imprudent to open-circuit a feedback line because of possible damage to other components.

Since in practice a wide variety of feedback arrangements are used, no standard rule exists for fault location in closed-loop systems. Procedures which may be used in one system may not be suitable in another system.

## Summary

In developing a logical approach to fault location the most useful main method to apply for general fault finding is that based on dynamic testing where each fault is approached with an 'open mind' and the technician systematically tests the individual stages or sections of the equipment using wherever possible the 'half-split' technique.

The use of reliability data and mechanical inspection techniques are, however, important additional methods which the technician will often need to apply in practice, particularly in field servicing when specialist dynamic testing equipment is not available.

CHAPTER TWO

# SCHEME FOR FAULT LOCATION

A LOGICAL APPROACH to fault location in series systems can be broken down into the following steps:

1 Functional checks
2 Initial tests
3 Dynamic tests
4 Static tests
5 Component replacement

### FUNCTIONAL CHECKS

When presented with an item of faulty equipment, the technician must first ascertain the actual fault symptoms, bearing in mind any reported symptoms. This will normally involve switching-on the equipment, if it is safe to do so, and carefully noting all symptoms or results as functional controls, push buttons or range switches, etc. are operated. For example, with a radio receiver this would imply checking for output while operating the volume control, tone control and waveband switch and tuning across each band. With a position-control servo system it would entail moving the reference potentiometer across its full range and noting the response of the motor and its load and, when fitted, operating the attenuator while observing the effect on the system.

Fault symptoms such as no-go, no output, low output or distorted output are usually self-evident, but equipment displaying intermittent operation or intermittent noise will require 'soak-testing' to confirm the symptoms. Sometimes faults develop only under high ambient temperature conditions, in which case the technician will be required to reproduce such an environment.

At this stage the technician will be establishing whether or not a fault does exist and, if so, the nature of the fault, to ensure that the equipment has failed – and not the user or operator!

In the interest of 'personal safety' a golden rule of servicing, whether it be

12  FAULT LOCATION IN ELECTRONIC EQUIPMENT

operational or diagnostic is always **check for youself**. For example, don't ask someone else to check that the mains switch is 'off', **check it yourself**.

## INITIAL TESTS

Having established the fault symptoms but before proceeding with the dynamic testing of the various stages or blocks in a system, it is often worthwhile making a few simple 'initial tests'. These may save unnecessary time and effort and may be instrumental in helping to locate the faulty area. Useful initial tests are:

(a) Physical inspection
(b) Current check
(c) D.C. supply or battery voltage check

(a) *Physical inspection*

This visual inspection should include checking the unit for obvious signs of component damage, e.g. broken or burnt components, severed leads, blown fuses, etc.

(b) *Current check*

This check, where applicable, can be most informative and involves measuring the d.c. current taken from the power supply under quiescent conditions, see Fig. 2.1.

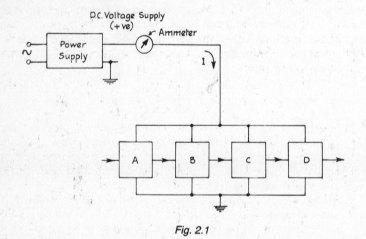

Fig. 2.1

Here an ammeter, set initially on a high range, is connected in series with the d.c. voltage supply to the stages A–D of the equipment under test so as to measure the current $I$ taken from the power supply. From the results of the measurement and comparing with the normal expected quiescent current, the following deductions can be made:

# SCHEME FOR FAULT LOCATION

(1) *Zero current* – indicating a discontinuity in the common supply line feed to the various stages perhaps due to a broken lead, cracked printed circuit or o/c fuse (if fitted), or the absence of a d.c. voltage output from the power supply.
(2) *Current too high* – indicating a lower resistance than normal to the common chassis line, due to, say, a leaky or s/c decoupling capacitor, incorrect biasing in any of the stages or internal break-down in any of the 'active' devices, e.g. thyristors, transistors, diodes, i.c.s, etc. If the current is quite high, carefully touch-test the active devices for excessive heat dissipation. Should the test be positive, internal break-down may be temporarily relieved by spraying the suspected component with an aerosol 'freezer' (which produces rapid cooling) while observing the ammeter reading.
(3) *Current too low* – indicating a higher resistance than normal to the common chassis line. This may be due to a common decoupling resistor being high in value, loss of bias on an active device, low or no d.c. gain in an active device or low voltage output from the power supply. Note that power stages when faulty will cause the greatest change in quiescent current.
(4) *Current correct* – indicating that the d.c. conditions in the various stages are most likely correct, thus pointing to a signal fault, e.g. o/c 'signal' path or unwanted attenuation in the 'signal' path.

(c) *D.C. supply or battery check*

Since failures often occur in the power section of equipment, it is useful to check the d.c. output voltage from the power supply at a point which is closest to the common feed line of the various stages. With battery operated equipment, a check of the 'on-load' battery voltage may save many unnecessary tests being made.

## DYNAMIC TESTS

Dynamic testing refers to the systematic 'signal' testing of the various stages, units or blocks within the system following the 'signal' path from beginning to end as outlined in Chapter 1 and when possible applying the 'half-split' technique. This will involve the use of a monitoring device and sometimes a 'signal injection' source.

The object at this point in the fault location procedure is to **isolate the faulty unit, stage or area of the equipment under test**. No gain or low gain between the input and output pins of an amplifying/processing device or loss of 'signal' in a particular area would indicate that the limit of dynamic testing had been reached.

## STATIC TESTS

Static tests are made when the limit of dynamic testing has been reached. If a fault is suspected between the input and output of an amplifying/

processing device, d.c. voltage measurements should be made on each pin of the device and comparison made with the service data. If the d.c. measurements are correct or lie within acceptable tolerances, an a.c. fault is indicated.

When the d.c. measurements are incorrect, deductions are then made as to the possible faulty component. The equipment should then be switched 'off' and resistance checks made on the suspected component, bearing in mind the circuit environment of the suspected component and the polarity of the ohmmeter test leads.

Static testing is considered in greater detail in Chapter 3.

## COMPONENT REPLACEMENT

Having removed the faulty component and carried out any further testing to confirm the results of the static tests, it must now be replaced. Before replacing it is necessary to know the parameters of the component and these are usually given in the service data. This applies whether the component is a simple resistor or complex i.c. It is essential that the replacement at least matches the manufacturer's specification or exceeds it.

There are some components which, in the interests of safety, must be replaced by the manufacturer's original type only. Such components are marked with the symbol ⓢ on the circuit diagram. Not only is the service engineer under a moral obligation to protect the user but also a legal one, and legal proceedings may be brought against a person ignoring the safety sign when replacing a component bearing the safety symbol.

CHAPTER THREE

# STATIC TESTS

WHEN THE LIMIT of dynamic testing has been reached and the 'signal' has been lost between the input and output of a discrete transistor stage or distortion or low gain is experienced, then voltage measurements should be made.

**TRANSISTOR VOLTAGE MEASUREMENTS**

**Potential divider**

Before actual measurements are considered, the action of a simple potential divider will be revised such as that used to provide the base potential of a transistor, see Fig. 3.1.

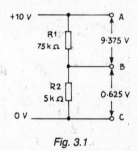

*Fig. 3.1*

From proportion it will be seen that the voltage across R2 is given by

$$\frac{5}{80} \times 10 = 0 \cdot 625 \text{V}$$

and that across R1 is given by

$$\frac{75}{80} \times 10 = 9 \cdot 375 \text{V}$$

These values are, of course, the theoretical voltages and the actual voltages

# 16 FAULT LOCATION IN ELECTRONIC EQUIPMENT

measured will depend upon the internal resistance of the voltmeter used and the resistor tolerances.

For example, if a voltmeter of 20kΩ/V sensitivity is used to measure the voltage across R2, then if set on its 2·5V range, the equivalent circuit will be as shown in Fig. 3.2(a) and the actual reading obtained will be 0·57V, an error of approximately 9%. With the meter set to its 10V range, the actual reading obtained will be 0·61V, an error of 2·4%.

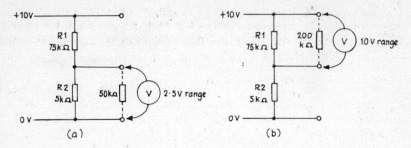

Fig. 3.2

With the voltmeter set on its 10V range and connected across R1, the equivalent circuit is as in Fig. 3.2(b). The actual voltage measurement obtained will be 9·16V, an error of approximately 2%. These meter loading errors must be taken into account when assessing the correct action of the potential divider.

Fig. 3.3(a) shows the effect of an open-circuited R1 on the voltage readings. As there is no current flow in R1 and R2 there will be no voltage drop across R2, thus the voltmeter will read zero when connected across R2. There is also no voltage drop across R1. However, when a voltmeter is connected across R1 as shown dotted, a reading will be obtained of value depending upon the internal resistance of the voltmeter relative to the resistance of R2. If a voltmeter with a sensitivity of 20kΩ/V is used and set on its 10V range, the reading obtained will be 9·75V (approximately 10V).

The effect of an open-circuited R2 is shown in Fig. 3.3(b). Again, since there is no current flow there is no voltage drop across R1 and the voltmeter

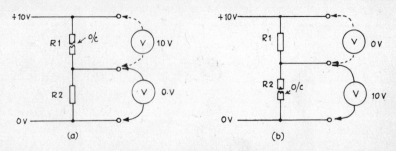

Fig. 3.3

## STATIC TESTS

will read zero when connected across R1. Also, there is no voltage drop across R2. However, when a voltmeter is connected across R2 a reading will be obtained of value depending upon the internal resistance of the voltmeter relative to the resistance of R1. If a 20k$\Omega$/V meter set on its 50V range is connected across R2 it will read 9·3V (approximately 10V).

**Single-stage Transistor Amplifier**

The circuit of Fig. 3.4 shows a class 'A' single-stage transistor amplifier suitable for use at audio frequencies. The voltages given are for normal operation and were measured with a 20k$\Omega$/V instrument set on its 10V range under no-signal conditions.

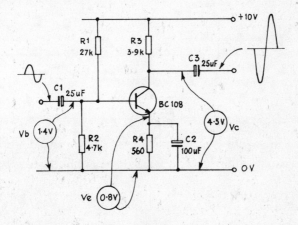

*Fig. 3.4*

It will be seen that the transistor has a forward bias of 0·6V ($V_b - V_e$) and that the transistor is passing an approximate collector current of

$$\frac{10 - 4·5}{3900} \times 10^3 \text{ mA} \approx 1·4 \text{mA}$$

The effect of a number of fault conditions will now be considered and the particular symptoms associated with each fault. **All voltages shown in the diagrams are positive with respect to the 0V line.**

# 18 FAULT LOCATION IN ELECTRONIC EQUIPMENT

**Fault (1)**
R1 open-circuit.

**Symptom**
No output signal.

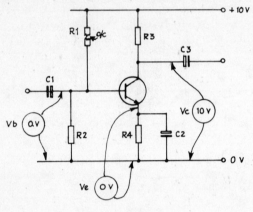

Fig. 3.5

**D.C. voltages**

|       | Fault | Normal |
|-------|-------|--------|
| $V_c$ | 10V   | 10V    |
| $V_b$ | 0V    | 1·4V   |
| $V_e$ | 0V    | 0·8V   |

**Explanation**

With R1 open-circuit there is no forward bias for the transistor, thus there is no base current. Therefore the base voltage is zero and there is no emitter current and no emitter voltage. As there is no collector current there is no voltage drop across R3 and in consequence the collector voltage is the same as the supply voltage, i.e. 10V.

# STATIC TESTS

**Fault (2)**
R2 open-circuit.

**Symptoms**
Severe distortion in output; negative-going half-cycles clipped.

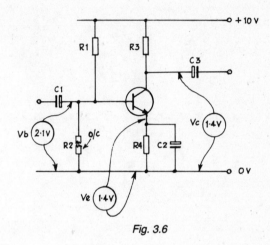

*Fig. 3.6*

**D.C. voltages**

|       | Fault | Normal |
|-------|-------|--------|
| $V_c$ | 1·4V  | 4·5V   |
| $V_b$ | 2·1V  | 1·4V   |
| $V_e$ | 1·4V  | 0·8V   |

**Explanation**

With R2 open-circuit the current that was flowing in R2 now tries to divert to the base. Thus the base current rises to a value that causes the collector current to saturate and for the collector voltage to 'bottom'. As the base current is larger, there is a greater base–emitter voltage drop (0·7V) and the emitter voltage must also rise. Since the collector voltage is 'bottomed', then, when a signal is applied the transistor will respond only to negative-going half-cycles at the base, taking the transistor out of saturation and resulting in only positive-going half-cycles at the collector.

**Fault (3)**
R3 open-circuit.

**Symptom**
No output signal.

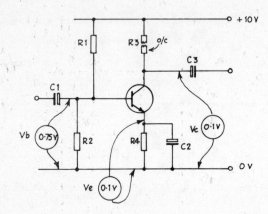

Fig. 3.7

**D.C. voltages**

|  | Fault | Normal |
|---|---|---|
| $V_c$ | 0·1V | 4·5V |
| $V_b$ | 0·75V | 1·4V |
| $V_e$ | 0·1V | 0·8V |

**Explanation**

As R3 is open-circuit there will be no collector current. The action of the potential divider R1, R2 still tries to maintain a fixed voltage on the base but the base voltage is lower than normal due to the increased voltage drop across R1 as a result of a larger-than-normal base current. As there is no collector current, the emitter current is very much less than normal and the emitter voltage is low (0·1V).

Because R3 is o/c it might be assumed that the collector voltage would be zero. However, when a voltmeter is connected it provides a high resistance path from the collector to the 0V line, causing the base–collector junction to become forward-biased. The voltage drop across the base–collector junction is 0·65V and the collector is 0·65V less than the voltage across R2, i.e. 0·1V.

# STATIC TESTS

**Fault (4)**
R4 open-circuit.

**Symptom**
No output signal.

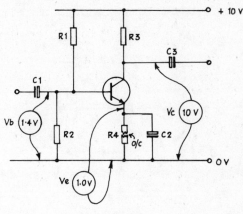

*Fig. 3.8*

**D.C. voltages**

|  | Fault | Normal |
|---|---|---|
| $V_c$ | 10V | 4·5V |
| $V_b$ | 1·4V | 1·4V |
| $V_e$ | 1·0V | 0·8V |

**Explanation**

With R4 o/c there will be no current through the transistor, i.e. no base, emitter or collector current. Since there is no collector current there is no voltage drop across R3, thus the collector voltage rises to the supply rail. Because the bleeder current in R1, R2 is very much larger than the normal base current, then, when the base current falls to zero, the rise in base voltage is barely noticeable.

When the voltmeter is connected to measure the emitter voltage, the emitter circuit is completed via the voltmeter. Due to the high resistance of the voltmeter, only a small emitter current flows but the emitter voltage is a little higher than normal.

# 22 FAULT LOCATION IN ELECTRONIC EQUIPMENT

**Fault (5)**
C2 open-circuit.

**Symptom**
Low gain.

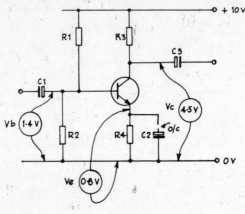

Fig. 3.9

**D.C. voltages**

|       | Fault | Normal |
|-------|-------|--------|
| $V_c$ | 4·5V  | 4·5V   |
| $V_b$ | 1·4V  | 1·4V   |
| $V_e$ | 0·8V  | 0·8V   |

**Explanation**

With C2 o/c the d.c. conditions are unaffected. The fault is identified by the 'low gain' symptom and arises since, with C2 o/c, a.c. signals will appear across R4 and introduce negative feedback. The voltage gain will then fall to a value given by R3/R4, i.e. a gain of approximately 7.

## STATIC TESTS

**Fault (6)**
C2 short-circuit.

**Symptom**
Very distorted output.

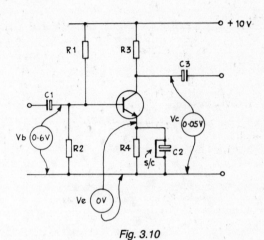

*Fig. 3.10*

**D.C. voltages**

|       | Fault  | Normal |
|-------|--------|--------|
| $V_c$ | 0·05V  | 4·5V   |
| $V_b$ | 0·7V   | 1·4V   |
| $V_e$ | 0V     | 0·8V   |

**Explanation**

As R4 is effectively shorted out, the emitter voltage is zero. The larger base voltage causes the transistor to conduct heavily and for the collector current to saturate. The collector current is limited to 10/3900 A = 2·6mA which prevents damage to the transistor. The collector voltage is 'bottomed' at 0·05V.

Because the collector voltage is 'bottomed', then, when a signal is applied the transistor will respond only to negative-going half-cycles, resulting in only positive-going half-cycles at the collector.

# 24 FAULT LOCATION IN ELECTRONIC EQUIPMENT

**Fault (7)**
Collector–base junction open-circuit.

**Symptom**
No output.

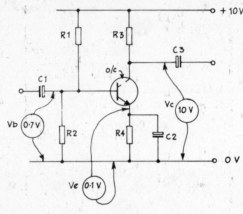

Fig. 3.11

**D.C. voltages**

|  | Fault | Normal |
|---|---|---|
| $V_c$ | 10V | 4·5V |
| $V_b$ | 0·7V | 1·4V |
| $V_e$ | 0·1V | 0·8V |

**Explanation**

Since the collector lead of the transistor is o/c there will be no collector current, thus the collector voltage will rise to that of the supply rail. The base–emitter junction acts like a forward-biased diode in a manner similar as for Fault (3) (R3 o/c).

# STATIC TESTS

**Fault (8)**
Collector–base junction short-circuit.

**Symptom**
No output.

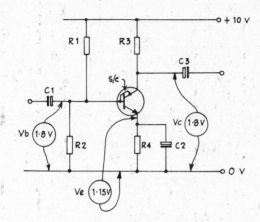

Fig. 3.12

**D.C. voltages**

|  | Fault | Normal |
|---|---|---|
| $V_c$ | 1·8V | 4·5V |
| $V_b$ | 1·8V | 1·4V |
| $V_e$ | 1·15V | 0·8V |

**Explanation**
    The indication of the s/c is given by the fact that collector and base voltages are the same. For this fault condition, the circuit is effectively reduced to the series combination of R3, the base–emitter diode and R4; the resistance of this path is much lower than that of R1, R2, which may be neglected. Thus the current in R4 is given by:

$$\frac{10 - 0·65}{R3 + R4} = \frac{10 - 0·65}{4·46k\Omega} = 2·01\text{mA}$$

Thus the emitter voltage will be $2·01 \times 10^3 \times 560 = 1·13$V (approximately 1·15V). The voltage on the base will be 0·65V higher than this (1·8V) which is sufficient to provide forward biase for the base–emitter diode.

# 26  FAULT LOCATION IN ELECTRONIC EQUIPMENT

**Fault (9)**
Collector–emitter short-circuit.

**Symptom**
No output.

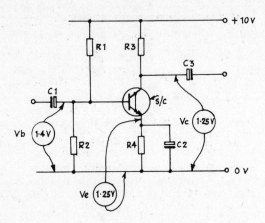

Fig. 3.13

**D.C. voltages**

|  | Fault | Normal |
|---|---|---|
| $V_c$ | 1·25V | 4·5V |
| $V_b$ | 1·4V | 1·4V |
| $V_e$ | 1·25V | 0·8V |

**Explanation**

The voltages on the collector and emitter are the same, indicating a s/c. The value of voltage is determined by the values of R3 and R4 which form a potential divider across the line supply voltage. The base voltage remains virtually unchanged, but because the emitter voltage has risen the base–emitter diode is cut-off.

# STATIC TESTS

**Fault (10)**
Base–emitter junction short-circuit.

**Symptom**
No output.

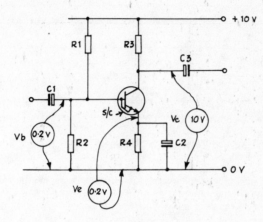

Fig. 3.14

**D.C. voltages**

|  | Fault | Normal |
|---|---|---|
| $V_c$ | 10V | 4·5V |
| $V_b$ | 0·2V | 1·4V |
| $V_e$ | 0·2V | 0·8V |

**Explanation**

The voltages at base and emitter being the same indicate a short circuit. The low voltage obtained is due to the fact that R4 is effectively in parallel with R2 under the fault condition. As the base and emitter are short-circuit there is no transistor action, causing the collector voltage to rise to the supply rail. When an input signal is applied it will be shorted-out via C2.

# 28 FAULT LOCATION IN ELECTRONIC EQUIPMENT

**Fault (11)**
Base–emitter junction open-circuit.

**Symptom**
No output.

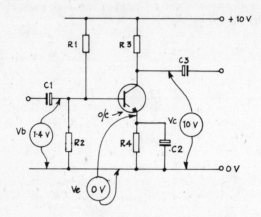

*Fig. 3.15*

**D.C. voltages**

|       | Fault | Normal |
|-------|-------|--------|
| $V_c$ | 10V   | 4·5V   |
| $V_b$ | 1·4V  | 1·4V   |
| $V_e$ | 0V    | 0·8V   |

**Explanation**

With the base–emitter junction o/c there is no current flow in the transistor, thus there is no voltage drop across R3 or R4. Therefore the emitter voltage is zero and the collector voltage rises to the supply rail. The base voltage remains approximately constant and is set by R1 and R2 values.

## COMPONENT TESTING

When the main logical procedures to isolate the fault to possibly one or two components have been carried out, the suspected components need to be tested to confirm whether or not replacement is required. In the interests of 'speed of repair' it is advantageous, where possible, to carry out the checks on the components while they are still soldered into circuit. Once a component has been diagnosed as being faulty, it can be removed from circuit and tested to confirm the original diagnosis. Testing components while still in circuit does, though, present special problems as will be highlighted in the examples that follow.

### Resistors

*Out-of-circuit*

Using an analogue or digital multimeter on the ohms range, compare the measured value with that given on the circuit diagram or that indicated by the resistor colour code, see Fig. 3.16.

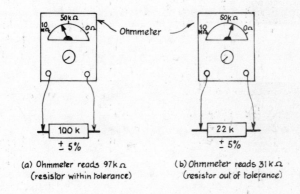

*Fig. 3.16*

The manufacturing tolerances, e.g. ±10%, ±5% or ±2% should be taken into account before deciding whether or not the resistor is faulty. Do not forget to 'zero' the ohms range before making the measurement. Also, take care not to grip the end of the resistor with your fingers as your body resistance will be in shunt with the resistor under test and lower the reading; this is most noticeable for resistors above 100kΩ in value.

*In-circuit*

The main problem with measuring resistors in circuit is that the measured value may not be the true value of the actual component due to alternative resistance paths within the circuit.

# 30  FAULT LOCATION IN ELECTRONIC EQUIPMENT

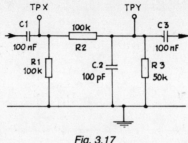

*Fig. 3.17*

Consider the circuit of Fig. 3.17 which lies in the path of a 'signal'. Suppose that during dynamic testing, the correct signal response is found at TPX but no signal is found at TPY. Thus the fault is confined to the area between TPX and TPY. For these particular symptoms there are **two** possible causes:

(a) an open circuit between TPX and TPY;
(b) a short circuit between TPY and chassis.

A useful general rule is **before checking across (or up) – check down**. Thus, after switching 'off' the equipment, connect the ohmmeter as shown in Fig. 3.18.

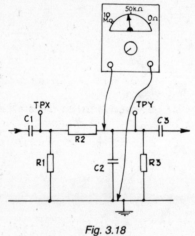

*Fig. 3.18*

Suppose that the resistance reading obtained is 51kΩ. This obviously rules out a short-circuit between TPY and chassis due to, say, C2 being s/c or a short existing between the printed circuit and chassis. A reading of 51kΩ points to an open-circuit between TPX and TPY since the resistance reading that should have been obtained would have been

$$\frac{R3(R1 + R2)}{R3 + (R1 + R2)}, \text{ i.e. approximately } 40k\Omega.$$

# STATIC TESTS

The reading of 51kΩ is that due to R3 alone, allowing for component tolerance.

The ohmmeter may be transferred to across R2, as shown in Fig. 3.19, where a reading of 152kΩ is obtained. This reading confirms an open-circuit

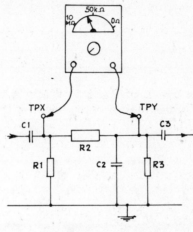

*Fig. 3.19*

between TPX and TPY since the reading is approximately that of R1 and R3 in series. If there were no open-circuit the resistance reading would have been

$$\frac{R2(R1 + R3)}{R2 + (R1 + R3)}$$, i.e. approximately 60kΩ.

Resistor R2 may now be removed from circuit and an out-of-circuit test made to confirm that R2 is open-circuit and not the wiring (printed circuit) either side of R2.

## Capacitors

*Out-of-circuit*

An ohmmeter, although limited in its use for capacitor testing, may be used to detect certain defects in capacitors, see Fig. 3.20.

When the dielectric of a capacitor is permanently 'broken-down', an ohmmeter will read a low resistance or zero ohms as in diagram (a) where a short-circuit is indicated.

A 'good' capacitor, i.e. one that is neither s/c or o/c will give the indication shown in Fig. 3.20(b). The capacitor should first be discharged by shorting its leads together and then be connected to an ohmmeter on a high-resistance range. If the capacitor is serviceable, the pointer will 'kick back' and then rise as the capacitor charges to the internal battery voltage of

32     FAULT LOCATION IN ELECTRONIC EQUIPMENT

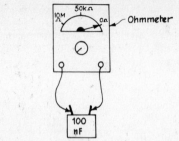

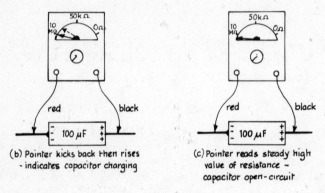

Fig. 3.20

the ohmmeter. The larger the value of capacitance the greater is the 'kick back'. Because the 'kick back' is very rapid with small value capacitors (less then $0{\cdot}01\mu F$), it is difficult to detect and can be mistaken for an open-circuit. For this reason and because the 'kick back' is only a rough guide to capacitance, a bridge or capacitance meter is more satisfactory for capacitor testing.

When testing electrolytic capacitors by this method, the capacitor should be correctly polarised by the internal battery of the ohmmeter. Note that for many multimeters, the **negative terminal** is connected to the **positive pole** of the internal battery.

Fig. 3.20(c) shows the effect for an open-circuit electrolytic capacitor where there is no 'kick back' and a high steady resistance (infinity) is obtained.

*In-circuit*

Short-circuits on capacitors while in-circuit can be detected readily by simple resistance checks using an ohmmeter. Open-circuit or change in capacitance value are more difficult to check while in-circuit, due to the effects of other components. Consider the circuit of Fig. 3.21 where a fault is known to exist.

# STATIC TESTS

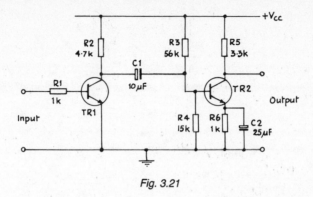

*Fig. 3.21*

Suppose that the correct signal response is found at TR1 collector but there is no response at TR2 base, indicating that the fault lies between TR1 collector and TR2 base. Possible faults are:

(a) C1 o/c;
(b) low resistance between the junction of R3/R4 and chassis.

The general rule should be applied – **before checking across – check down**. 'Checking-down' may be achieved either by switching off and making a resistance measurement between the junction of R3/R4 and chassis or by making a voltage measurement at TR2 base. If the correct base voltage is found it may be assumed that the suspected low resistance does not exist. This leaves an open-circuit C1 as the possible fault.

If a resistance check is made across C1, then the resistance reading obtained will be that of R2 and R3 in series, i.e. approximately 61k$\Omega$. When C1 is serviceable, a 'kick-back' should be observed before the steady reading of 61k$\Omega$ is obtained. If C1 is open-circuit there will be no initial 'kick-back'. However, with a coupling capacitor of smaller value, say 0·01µF, it will not be possible to detect the 'kick-back'. Thus the only satisfactory way of checking C1 is to remove it from circuit for testing. Alternatively, the equipment may be switched off and C1 bridged with another capacitor of the correct value and polarity to see if the correct response is obtained at TR2 base when the equipment is tested dynamically on switching-on again.

## Inductors

A common fault with inductors such as tuning coils, relay coils, chokes and transformer windings is that they go open-circuit, which can be detected readily with an ohmmeter. Inductors may develop shorted turns, which is a more difficult fault to detect with an ohmmeter as the change in d.c. resistance can be quite small. With transformers or other inductors with a core there may be a break-down in the insulation between the winding and the core. This may show up as a low resistance on an ohmmeter but often the break-down occurs only under normal operating voltage stress. In this case a

Megger may be used to detect the fault – ensuring that the test voltage does not exceed the normal operating voltage.

Break-down in the insulation between the winding (primary or secondary) and the core of a transformer or shorted turns may result in the transformer core getting excessively hot which can be detected readily and may be accompanied by the smell of burning or the melting of insulation.

*In-circuit*

Open-circuits in inductors may be detected easily while in-circuit as illustrated by the examples in Fig. 3.22.

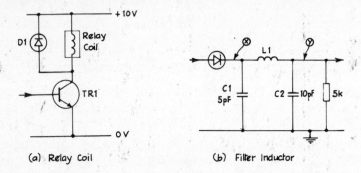

Fig. 3.22

Suppose that the relay coil of Fig. 3.22(a) is suspected of being faulty. Under normal operating conditions the collector voltage of TR1 will be low when the transistor is 'on' and equal to the supply rail when TR1 is 'off'. If the collector voltage reads zero in the 'off' condition, the relay coil is probably o/c (assuming that the supply rail voltage is present). Thus if an ohmmeter is connected across the coil after switching 'off' the equipment, it will read low resistance in one direction and high resistance on reversing the leads if o/c. The low and high resistance readings are the forward and reverse resistance of D1 (protection diode). If the coil is not o/c a low resistance reading will be obtained in either direction.

Consider now the demodulator circuit of Fig. 3.22(b) and suppose that, during dynamic testing, the correct signal response is obtained at $X$ but there is no response at $Y$. The fault could be caused by an open-circuit L1 or a short-circuit C2. After switching-off, 'check down' first by connecting an ohmmeter across C2 (a 5kΩ reading should be obtained). If this is satisfactory, connect the ohmmeter across L1; it will read infinity is L1 is o/c or low resistance if the coil is not o/c.

**Semiconductor Diodes**

Semiconductor diodes of both signal and power types are characterised by a low forward resistance in one direction when forward biased and a high

# STATIC TESTS 35

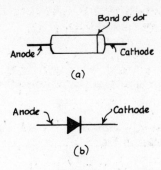

Fig. 3.23

resistance in the other direction when reverse biased. The diagrams of Fig. 3.23 show a p–n diode and its circuit symbol.

*Out-of-circuit*

The main fault that occurs with diodes is that the junction breaks down and goes short-circuit, or internal connections burn out producing an open-circuit. These faults may be detected readily using an ohmmeter, see Fig. 3.24.

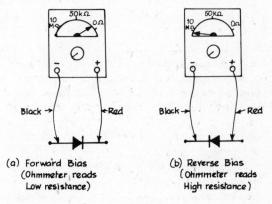

Fig. 3.24

Fig. 3.24(a) shows the normal low resistance that should be obtained when the diode is forward biased by the ohmmeter battery, and Fig. 3.24(b) shows the normal high resistance when the diode is reverse biased. Typical resistances for working diodes are given in the following table.

| Type | Forward resistance | Reverse resistance |
|---|---|---|
| Signal diode (Ge) | 80Ω | 3MΩ |
| Signal diode (Si) | 20Ω | Infinity |
| Power diode (Si) | 15Ω | Infinity |

# 36 FAULT LOCATION IN ELECTRONIC EQUIPMENT

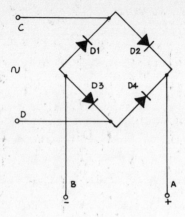

*Fig. 3.25*

Diodes that measure the same resistance in the forward and reverse directions are faulty.

Consider the out-of-circuit testing of the bridge rectifier unit of Fig. 3.25. With 1A silicon rectifier diodes one would expect a low forward resistance of about 15Ω, and a very high reverse resistance (infinity) for each diode. Thus, if all diodes are working, the following resistance readings would be obtained between the lettered points on the diagram:

**All Diodes Working**

| Ohmmeter between | Resistance reading | Reason |
|---|---|---|
| A–B | | |
| A positive w.r.t. B | Infinity | All diodes reverse biased |
| A negative w.r.t. B | 15Ω | All diodes forward biased |
| C–D | | |
| C positive w.r.t. D | Infinity | Ohmmeter reads the resistance of D1 and D3 back-to-back in parallel with D2 and D4 back-to-back |
| C negative w.r.t. D | Infinity | |
| C–B | | |
| B positive w.r.t. C | 15Ω | Ohmmeter effectively reads forward and reverse resistance of D1 |
| B negative w.r.t. C | Infinity | |
| C–A | | |
| C positive w.r.t. A | 15Ω | Ohmmeter effectively reads forward and reverse resistance of D2 |
| C negative w.r.t. A | Infinity | |

## STATIC TESTS

| Ohmmeter between B–D | Resistance reading | Reason |
|---|---|---|
| B positive w.r.t. D | 15Ω | Ohmmeter effectively reads forward and reverse resistance of D3 |
| B negative w.r.t. D | Infinity | |

| D–A | | |
|---|---|---|
| D positive w.r.t. A | 15Ω | Ohmmeter effectively reads forward and reverse resistance of D4 |
| D negative w.r.t. A | Infinity | |

Consider now the effect of the following faults on the resistance readings:

### D1 Short Circuit

| Ohmmeter between A–B | Resistance reading | Reason |
|---|---|---|
| A positive w.r.t. B | Infinity | Ohmmeter effectively reads reverse or forward resistance of series D3 and D4 in parallel with D2 |
| A negative w.r.t. B | 10Ω | |

| C–D | | |
|---|---|---|
| C positive w.r.t. D | 15Ω | Ohmmeter effectively reads forward and reverse resistance of D3 |
| C negative w.r.t. D | Infinity | |

| C–B | | |
|---|---|---|
| C positive w.r.t. B | 0 | Ohmmeter reads short-circuit of D1 |
| C negative w.r.t. B | 0 | |

### D1 Open Circuit

| Ohmmeter between A–B | Resistance reading | Reason |
|---|---|---|
| A positive w.r.t. B | Infinity | Ohmmeter reads reverse or forward resistances of D3 and D4 in series |
| A negative w.r.t. B | 30Ω | |

| C–D | | |
|---|---|---|
| C positive w.r.t. D | Infinity | Ohmmeter reads resultant resistance of D2 and D4 back-to-back |
| C negative w.r.t. D | Infinity | |

| C–B | | |
|---|---|---|
| C positive w.r.t. B | Infinity | Ohmmeter effectively reads reverse resistance of D4, D3 or D2 |
| C negative w.r.t. B | Infinity | |

*In-cicuit*

When measuring the forward and reverse resistances of an in-circuit diode, parallel resistance paths formed by other circuit components must be taken into account, see Fig. 3.26.

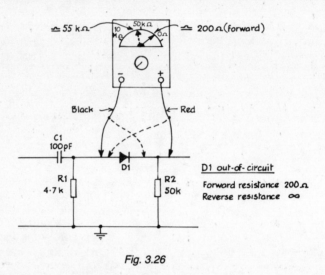

Fig. 3.26

Suppose that D1 is suspected of being faulty and typical out-of-circuit resistances for the diode are 200Ω (forward) and infinity (reverse). If the diode is not faulty, the resistance obtained when forward-biased by the ohmmeter battery would be

$$\frac{200 \times (50{,}000 + 4{,}700)}{200 + (50{,}000 + 4{,}700)} \approx 200\Omega$$

but that obtained in the reverse direction would be

$$4{,}700 + 50{,}000 \approx 55\text{k}\Omega$$

An open-circuit diode would produce readings of $4{,}700 + 50{,}000 \approx 55\text{k}\Omega$ regardless of test-lead polarity and a short-circuit diode would give zero ohms regardless of test-lead polarity.

**Transistors**

Bipolar transistors may be considered as consisting of two diodes connected back-to-back, as illustrated in Fig. 3.27.

As with p–n diodes, the main fault that occurs with transistors is that the junctions break down and go short-circuit or their internal connections burn out and go open-circuit. These defects may be tested with an ohmmeter although, except for short-circuits, the results are not always conclusive, in which case a special 'transistor checker' or 'curve tracer' may be required.

# STATIC TESTS

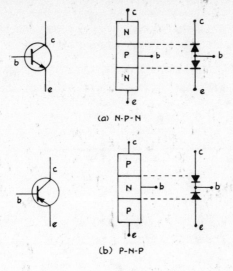

Fig. 3.27

## Out-of-circuit

The diagram of Fig. 3.28 shows the use of an ohmmeter for checking the base–emitter junction of a transistor. If the transistor is serviceable, a low resistance will be obtained in the forward direction and a high resistance in the reverse direction. A short-circuited junction would read zero ohms and an open-circuit junction would read infinity, regardless of the ohmmeter polarity.

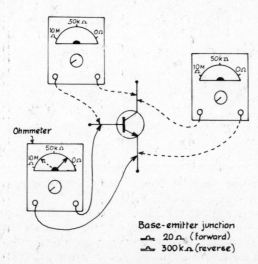

Fig. 3.28

# 40 FAULT LOCATION IN ELECTRONIC EQUIPMENT

The table below gives some typical resistances for serviceable transistors but it should be noted that the actual resistance values obtained will depend upon the test voltage supplied by the ohmmeter battery.

| Transistor | Ohmmeter between | Resistance reading |
|---|---|---|
| BC108 (Si) 300mW | Emitter–Base | 20Ω<br>300kΩ |
|  | Base–Collector | 15Ω<br>Infinity |
|  | Collector–Emitter | 60kΩ<br>Infinity |
| 2N3055 (Si) 115W | Emitter–Base | 10Ω<br>2MΩ |
|  | Base–Collector | 10Ω<br>Infinity |
|  | Collector–Emitter | 2MΩ<br>Infinity |

*In-circuit*

When making resistance checks on a transistor while it is still connected in-circuit, it is necessary to make allowances for other resistance paths in parallel with the junction under test, see Fig. 3.29.

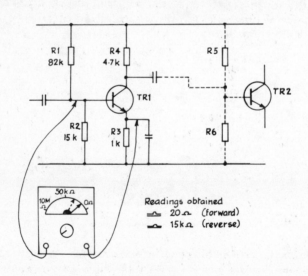

*Fig. 3.29*

Suppose that the base–emitter junction resistance is to be checked and that the expected out-of-circuit resistances are 20Ω (forward) and 300kΩ (reverse). It will be noted that the series combination of resistors R2 and R3 are effectively in parallel with the base–emitter junction, thus the actual resistance values obtained would be

$$\frac{20 \times 16\,000}{20 + 16\,000} \approx 19.7\Omega \text{ (forward)}$$

and

$$\frac{300\,000 \times 16\,000}{300\,000 + 16\,000} \approx 15\text{k}\Omega \text{ (reverse)}$$

In this particular case, due to the effects of R2 and R3, the forward resistance appears about normal but the reverse resistance is considerably lower. Here, only the components present in the single transistor stage of TR1 have been taken into account. In practice, the effects of resistance loading from other stages must be allowed for such as R5, R6 and TR2 junctions.

A short-circuited base–emitter junction would read zero ohms regardless of meter polarity and the effects of R2, R3. An open-circuit base–emitter junction would read 16kΩ (R2 + R3) regardless of meter polarity and ignoring the effects of R5, R6.

**Silicon Controlled Rectifiers**

Silicon controlled rectifiers are four-layer semiconductor devices, see Fig. 3.30. They are commonly used in a.c. power applications such as lamp dimmers, motor speed control, temperature control and inverters.

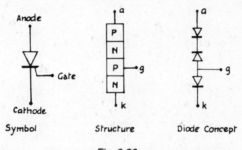

*Fig. 3.30*

*Out-of-circuit*

The p-n junctions of an SCR can be tested to a certain degree using simple resistance checks. The gate-to-cathode junction can be tested in the same way as the base-to-emitter junction of an ordinary transistor, see Fig. 3.31. In the forward-conducting direction a low reading of typically 1kΩ will be obtained but in the reverse direction a very high resistance (infinity) will be present.

A very high resistance will be present between the gate and anode

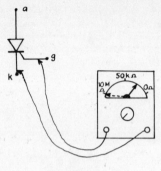

Fig. 3.31

regardless of the polarity of the test-meter leads. These out-of-circuit resistance checks are somewhat limited in that faults such as the anode being open-circuit, the device breaking-over at high voltage or the device failing to remain conducting once triggered may not show up.

*In-circuit*

In many power control circuits the SCR is fed on its gate with a series of short-duration pulses. To ensure that the pulses are present at the gate and at the same time to see what is happening at the anode, a dual-trace c.r.o. may be connected as shown in Fig. 3.32. The waveforms given in the diagram are for normal operation of the SCR.

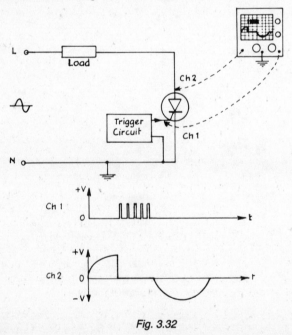

Fig. 3.32

# STATIC TESTS

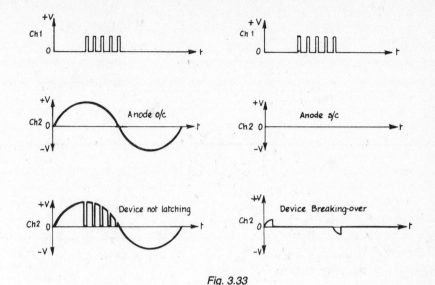

*Fig. 3.33*

The waveforms given in Fig. 3.33 show the expected results for various fault conditions occurring within the SCR.

## Integrated Circuits

Integrated circuits contain complete electronic circuits fabricated on to a single silicon chip. Because with many i.c.s there is no complete circuit diagram available for the device and also because a large number of internal components have no direct connection to the pins of the i.c., out-of-circuit testing is virtually meaningless. However, the function of the i.c. is usually known and input/output signal pins may be identified from an understanding of the system into which it is incorporated or by reference to manufacturer's data. Thus i.c.s have to be tested in-circuit where all external signals and supply voltages are present.

*In-circuit*

Consider the linear amplifier of Fig. 3.34 and suppose that during dynamic testing the correct signal response is found at pin 1 but no output is observed on pin 4. It is quite possible that the i.c. is faulty, but first it should be established as a general rule that the i.c. is receiving its correct d.c. potentials. For the i.c. shown this would entail:

(a) Check for the presence of d.c. voltage feeds to pins 2 and 4.
(b) Check that pin 5 is at the 0V line potential.

This may be achieved by connecting the voltmeter in turn between pins 2 and 5 and between pins 4 and 5. Always **test on the pins of the i.c.** rather than

# 44 FAULT LOCATION IN ELECTRONIC EQUIPMENT

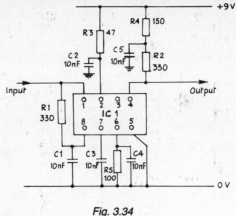

Fig. 3.34

on external component leads. According to the results obtained, the following deductions can be made:

**Zero voltage on pins 2 and 4**
Possible faults are defective positive supply line feed to the i.c. or an o/c between the 0V line and pin 5.

**Voltage on pin 2 correct, no voltage on pin 4**
Possible faults are R2 or R4 o/c, C5 s/c, i.c. internal s/c or s/c on external printed circuit connected to pin 4.

**Voltage on pin 4 correct, no voltage on pin 2**
Possible faults are R3 o/c, C2 s/c or i.c. internal s/c.

**Voltage on pin 4 correct, pin 2 at supply voltage** (or vice-versa)
Internal disconnection in i.c. or R5 (an external biasing resistor) o/c.

Another example is considered in Fig. 3.35 where a 555 timer i.c. is used as a monostable oscillator. The oscillator is triggered by negative-going

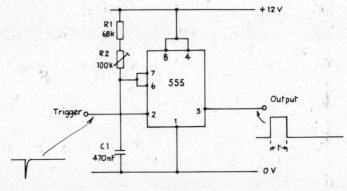

Fig. 3.35

## STATIC TESTS

pulses on pin 2, producing at the output on pin 3 positive-going pulses having a duration of $t$ seconds determined by the time constant R1, R2 and C1. Suppose, during dynamic testing with a c.r.o., that the correct trigger is found on pin 2 but there is no output on pin 3.

To eliminate external factors as the cause of the fault, the following checks should be made before replacing the i.c.:

(a) Check for the supply line voltage on pins 8 and 4.
(b) Check that pin 1 is at the 0V line potential.
(c) Check that pins 6 and 7 are at a positive potential.

A voltmeter connected in turn between pins 8 and 1 and between pins 4 and 1 will satisfy checks (a) and (c). If pins 6 and 7 are found to be at 0V, possible faults are C1 s/c, R1 or R2 o/c or i.c. internal s/c. If pins 6 and 7 are found to be at +12V possible faults are C1 o/c or i.c. internal fault.

Some general observation points about i.c.s are:

(a) Is the i.c. inserted into circuit the right way round? See Fig. 3.36. The orientation of the i.c. socket or p.c.b. holes can usually be found by

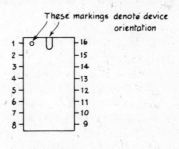

Fig. 3.36

identifying the connections to the supply voltage rails. I.C. orientation must be checked after replacement and before switching-on, otherwise the i.c. may be damaged permanently.
(b) Is the i.c. burned or cracked? Many i.c.s which have been destroyed by overheating show visual signs of being faulty.
(c) Are the pins of the i.c. in good condition? I.C.s that have been inserted into sockets sometimes have one or more pins bent underneath or even missing. Equipment that has been operated in dusty, damp or contaminated conditions may have corroded i.c. contacts.

CHAPTER FOUR

# FAULT TRACING IN LOGIC SYSTEMS

FAULT LOCATION IN digital circuits is concerned with the detection of the logic values of 1 and 0 at voltage levels appropriate to the type of logic family used, for example TTL, CMOS, ECL, etc. The logic levels may be static in some parts of the circuit for a period of time, changing only when input conditions alter, as with logic gates. In other areas of a system the logic levels may be in the form of repetitive pulses, such as clock pulses or counter outputs or brief single pulses as with strobe pulses.

Useful fault location aids include an ordinary c.r.o. for the display of **repetitive-type waveforms**, a logic probe for the detection of short-duration pulses and static logic levels, and a logic pulser for use as a pulse-generating source. Detection of non-repetitive waveforms such as blocks of binary data of different binary coding require the use of a special c.r.o. with digital storage facility.

### USE OF LOGIC PROBE

A logic probe is probably the most useful aid and can be employed for tracing the majority of faults in digital circuits. Typically red and green l.e.d.s are used to indicate the logic 1 and logic 0 levels respectively and its compactness makes it an ideal portable servicing tool, see Fig. 4.1.

This probe may be powered from the logic circuit under test or from an external power source, but in this case the negative line must be commoned

Fig. 4.1

# FAULT TRACING IN LOGIC SYSTEMS

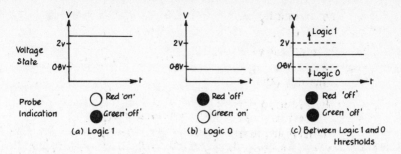

*Fig. 4.2*

with the logic system's earth. Some probes such as TTL/CMOS are dual purpose, but others may be designed for use with one logic family only. The logic state on a relevant i.c. pin can be detected by applying the probe and noting the response of the two l.e.d.s.

For TTL, the logic probe has its thresholds set to 0·8V (logic 0) and 2V (logic 1). The static logic circuit states indicated by the probe are shown in Fig. 4.2. When the logic level is at or above the logic 1 threshold of 2V, the red l.e.d. is 'on' and the green l.e.d. is 'off' as in Fig. 4.2(a). If the logic level is at or below the logic 0 threshold of 0·8V, the red l.e.d. is 'off' and the green l.e.d. is 'on' as in Fig. 4.2(b). Should, however, the voltage level in a circuit lie between the logic 1 and 0 thresholds as in Fig. 4.2(c), both l.e.d. will be 'off'; a display condition which applies also if the point under test is o/c (floating).

The diagrams of Fig. 4.3 show the types of fault that can occur at gate outputs which may be detected with a logic probe.

If the gate output is permanently stuck at logic 0 as in Fig. 4.3(a) irrespective of the logic conditions at the input (which may be checked with the probe), possible faults are:

(a)    + 5V supply line o/c internally or externally.
(b)    Internal transistor short.

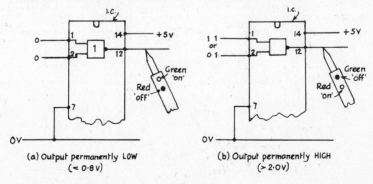

*Fig. 4.3*

# 48 FAULT LOCATION IN ELECTRONIC EQUIPMENT

Should the gate output be permanently stuck at logic 1 as in Fig. 4.3(b), possible faults are:

(a) 0V line o/c internally or externally.
(b) Internal transistor short.
(c) Internal transistor o/c.

When using the probe to determine input/output logic levels of an i.c. always **check on the actual pins of the i.c.** rather than on the printed circuit lines or board connections, to avoid missing o/c lines. Note that the probe may be used to check for the presence of a voltage on the + 5V supply line (but not the magnitude) and the 0V level which is useful for quickly eliminating o/c feeds.

## Pulse Detection

Logic probes are not restricted to static level testing. Light-emitting diodes are high-speed devices with light rise and fall times of less than 100ns. However, to be able to detect short-duration pulses visually it is necessary to lengthen the pulses. This may be achieved by incorporating into the probe a transient detector and, say, a monostable oscillator. When the narrow pulse is detected, the transient may be used to trigger the monostable which provides an output pulse of about 100ms duration. Thus if a probe is connected to a point which is at logic 0 and experiences a positive-going pulse of duration as little as 30ns, the green l.e.d. is pulsed 'off' for about 100ms, which is long enough for an observer to detect. Short-duration, negative-going pulses may also be detected causing the red l.e.d. to flash 'off' for about 100ms.

The diagrams of Fig. 4.4 show the expected l.e.d. indications for pulses of various durations and repetition frequencies. Thus a logic probe is useful for detecting clock pulses up to about 20MHz, counter outputs, strobe pulses, etc. The full versatility of a particular probe can be achieved only by studying the manufacturer's instruction sheet carefully.

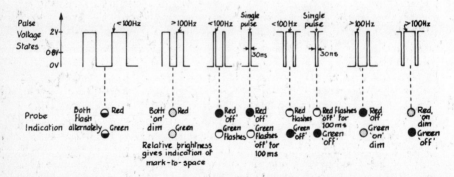

Fig. 4.4

# FAULT TRACING IN LOGIC SYSTEMS

## USE OF LOGIC PULSER

A logic pulser is essentially a compact pulse generator and in appearance is similar to a logic probe, see Fig. 4.5.

Pulsers normally have tri-state output which means that they can pulse **high** or **low** but present a high impedance when not operating. This feature permits the pulser to be left connected without interfering with the logic circuit under test. The pulser is powered from the circuit under test via its clip leads and typically will deliver either a single pulse, a group of four pulses or a continuous train of pulses, the operation being selected by a three-position slide switch. To prevent excessive dissipation in the device under test, the pulses are of short duration (0·8μs for TTL and 1·8μs for CMOS) at a repetition rate of 1kHz thus maintaining a low mean power dissipation.

*Fig. 4.5*

To use the pulser the tip is connected to the logic node or gate input, the appropriate option selected and the push-button operated. The pulser then automatically drives the circuit connected to it, producing a brief change in logic state which may be detected with a logic probe or c.r.o. The pulses produced are capable of sinking of sourcing up to about 0·5A which ensures sufficient current to override an i.c. output in either the **low** or **high** state. Typically a pulser will have a fan-out of 10, i.e. it will drive ten TTL gates simultaneously.

When considering the operation of the logic probe at gate outputs it was assumed that suitable logic levels were present at the gate inputs. Sometimes a gate input is present in a logic system only when, for example, a transducer is activated and this may occur infrequently. Thus the pulser may be used to replace the transducer source enabling circuit testing to be carried out whenever desired.

The pulser output may also be used to provide clocking of a counter using either the single pulse, multiple pulse or continuous pulse output. Occasionally in a counting system, false counts occur which may be due to a defective clocking signal, defective counter or the pick-up of noise pulses. By inhibiting the normal clocking signal and substituting the pulser output instead, the counter operation may be checked, see Fig. 4.6. If the single pulse or 4-pulse output is selected, then each time the pulser is operated a known number of pulses is applied to the counter. This 'slow clocking' of the counter enables faults to be detected more easily.

# 50  FAULT LOCATION IN ELECTRONIC EQUIPMENT

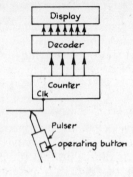

*Fig. 4.6*

Additional uses for the pulser include triggering of monostable oscillators, clock pulses for S–R bistables, strobe pulse for gate operation and the checking of bus and printed circuit line continuity in conjunction with a logic probe.

## SYSTEM FAULT LOCATION TECHNIQUES

### Digital Control System

Consider the logic circuit of Fig. 4.7 forming part of a control system where the repetitive outputs from 1, 2 and 3 are used to actuate certain functions within the system. The 3-bit binary counter counts the 10kHz clock pulses applied to its clock input. Inspection of the truth table in Fig. 4.7(b) shows the logic state of the outputs after the application of each clock pulse. It will be seen that the counter counts up to decimal 5 and on the sixth clock pulse the counter is reset by the action of the NAND gate G10.

Because the counter is clocked continuously, the repetitive output waveforms shown in Fig. 4.8 will be obtained having time durations that may be deduced from the truth table and the p.r.f. of the clock pulses; note that the periodic time of the clock pulses will be

$$\frac{1}{10\,000}\,\text{s} = 100\mu\text{s}.$$

A c.r.o. is probably the most useful main aid to fault location for this particular system. By 'scoping' outputs 1, 2 and 3 in turn, the correct/incorrect response may be ascertained. If all outputs are missing or all are incorrect, the fault is most likely confined to the clock pulse generator, the counter or G10. The clock pulse generator and counter may be eliminated by 'scoping' the clock input and the $Q_A$, $Q_B$ and $Q_C$ outputs of the counter, checking for the correct amplitude and duration of the waveforms. The most likely fault with the G10 is its output being permanently stuck at logic 0 thus keeping the counter permanently in the reset condition. 'Slow clocking' of the counter using a logic pulser in place of

# FAULT TRACING IN LOGIC SYSTEMS

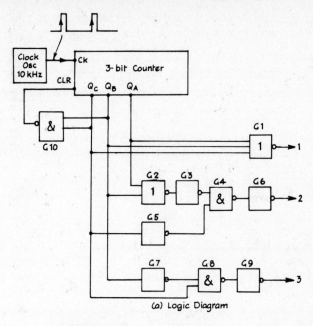

(a) Logic Diagram

| Counter Output | | | Outputs | | |
|---|---|---|---|---|---|
| $Q_C$ | $Q_B$ | $Q_A$ | 3 | 2 | 1 |
| 0 | 0 | 0 | 0 | 0 | 1 |
| 0 | 0 | 1 | 0 | 1 | 0 |
| 0 | 1 | 0 | 0 | 1 | 0 |
| 0 | 1 | 1 | 0 | 1 | 0 |
| 1 | 0 | 0 | 1 | 0 | 0 |
| 1 | 0 | 1 | 1 | 0 | 0 |
| 1 | 1 | 0 | Counter Resets | | |

(b) Truth Table

*Fig. 4.7*

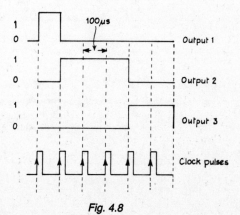

*Fig. 4.8*

## 52  FAULT LOCATION IN ELECTRONIC EQUIPMENT

the normal clocking signal may assist in eliminating false counts due to the pick-up of noise or intermittent clocking, as mentioned previously.

If outputs 1 and 3 are normal but output 2 is absent or incorrect, the fault is confined to gates G2–G6. The logic state of the intermediate gate outputs may be established by producing a truth table as shown in Fig. 4.9.

| $Q_C$ | $Q_B$ | $Q_A$ | G2 O/P | G3 O/P | G5 O/P | G4 O/P |
|---|---|---|---|---|---|---|
| 0 | 0 | 0 | 1 | 0 | 1 | 1 |
| 0 | 0 | 1 | 0 | 1 | 1 | 0 |
| 0 | 1 | 0 | 0 | 1 | 1 | 0 |
| 0 | 1 | 1 | 0 | 1 | 1 | 0 |
| 1 | 0 | 0 | 1 | 0 | 0 | 1 |
| 1 | 0 | 1 | 0 | 1 | 0 | 1 |

Fig. 4.9

The waveforms for the intermediate gate outputs (deduced from the truth table) are shown in Fig. 4.10 and these may be used for comparison with the c.r.o. displays obtained during fault location on the intermediate gate outputs.

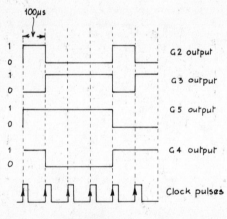

Fig. 4.10

### Binary Code Generator

Another example is considered in Fig. 4.11 which illustrates the basic operation of a code generator. This type of arrangement forms the basis of a keyboard encoder, where for each key that is depressed a particular binary code is generated by the ROM.

In the arrangement shown, switches A and B which may be at logic 1 or logic 0 provide an equivalent to four keyboard switches. Gates G1–G8 form a 2–4 line decoder enabling the selection of one of four 8-bit words stored in the diode matrix ROM. The four 8-bit words shown in the table are the

# FAULT TRACING IN LOGIC SYSTEMS

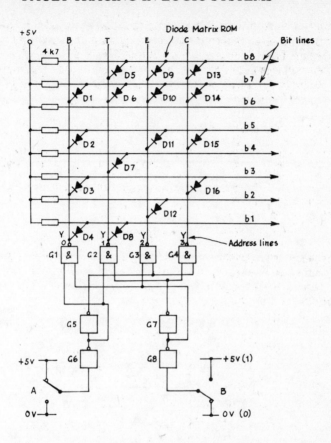

| A | B | b8 | b7 | b6 | b5 | b4 | b3 | b2 | b1 |
|---|---|----|----|----|----|----|----|----|----|
| 0 | 0 | 1  | 0  | 1  | 0  | 1  | 0  | 1  | 0  |
| 0 | 1 | 0  | 0  | 1  | 1  | 0  | 1  | 1  | 0  |
| 1 | 0 | 0  | 0  | 1  | 0  | 1  | 1  | 0  | 1  |
| 1 | 1 | 0  | 0  | 1  | 0  | 1  | 0  | 1  | 1  |

*Fig. 4.11*

ASCII codes for the letters BTEC where bit 8 is the parity bit (even parity). A particular word is selected by taking **one** of the Y address lines **low**, all other address lines remaining **high**.

The most appropriate method of fault location on this type of system is a 'software' approach, namely the use of truth tables. As the logic states on the bit lines are static and only change when switches A and B are altered, a c.r.o. is **not** a suitable aid to fault detection. The logic state on the bit lines is best checked with a logic probe, testing for the correct logic state on the eight bit lines for each of the four combinations of switches A and B.

## 54  FAULT LOCATION IN ELECTRONIC EQUIPMENT

Faults may occur in the ROM, G1–G8, switches A and B or their supplies. In the discrete component ROM shown it is possible to check for faults such as a diode o/c or s/c, but in an i.c. ROM this will not be possible.

Assume that the d.c. supplies have been established and that the bit line coding is incorrect for a single setting of switches A and B. It is first necessary to establish that the appropriate Y address line is **low**, all other address lines remaining **high**, see truth table of Fig. 4.12.

| Inputs | | | | | | Outputs | | | |
| --- | --- | --- | --- | --- | --- | --- | --- | --- | --- |
| | | | | | | $Y_0$ | $Y_1$ | $Y_2$ | $Y_3$ |
| A | B | G6 | G5 | G8 | G7 | G1 | G2 | G3 | G4 |
| 0 | 0 | 1 | 0 | 1 | 0 | 0 | 1 | 1 | 1 |
| 0 | 1 | 1 | 0 | 0 | 1 | 1 | 0 | 1 | 1 |
| 1 | 0 | 0 | 1 | 1 | 0 | 1 | 1 | 0 | 1 |
| 1 | 1 | 0 | 1 | 0 | 1 | 1 | 1 | 1 | 0 |

Fig. 4.12

Consider the following fault conditions and conclusions:

(1) For switch positions A = 0 and B = 1, b4 incorrect (**high** instead of a **low**) – all other bit lines correct.
On checking the logic state of the Y address lines with a logic probe it was noted that: $Y_0 = 1$, $Y_1 = 0$, $Y_2 = 1$ and $Y_3 = 1$.
Since the address lines are correct, the fault is confined to the ROM and the most likely cause is D7 o/c which may be confirmed by a resistance check.

(2) For switch position A = 0 and B = 0, bits b1 to b8 all **high**. For other switch positions the ROM output coding is correct.
On checking the logic state of the Y address lines with a logic probe it was noted that: $Y_0 = 1$, $Y_1 = 1$, $Y_2 = 1$ and $Y_3 = 1$.
Clearly the $Y_0$ address line should be **low**. Action: Check inputs to G1 – both found to be **high** (correct, see Fig. 4.12). Conclusion is that G1 is faulty – output permanently stuck at logic 1.

(3) For switch positions A = 1 and B = 0, bits b8 to b1 are 00101011 (incorrect). The ROM output coding is also incorrect for A = 0 and B = 0.
With switches at positions A = 1 and B = 0, check the logic state with the probe on the Y address lines. Results obtained: $Y_0 = 1$, $Y_1 = 1$, $Y_2 = 1$ and $Y_3 = 0$.
Check inputs to G3: Input from G5 **high** (correct) but input from G8 **low** (incorrect, see Fig. 4.12). Check input to G8: **low** (correct).
Conclusion: G8 output permanently **low**.

In the above manner, by adopting a logical approach with the aid of a truth table, faults can usually be isolated quickly to a single gate or component.

CHAPTER FIVE

# FAULT LOCATION EXERCISES

THESE QUESTIONS MAY be used to test the assimilation of fault finding procedures outlined in the previous chapters and the application of logical reasoning to faults in simple circuits and systems. Solutions are given at the end of the chapter.

Questions 1–5 refer to the series-connected amplifying chain of Fig. 5.1.

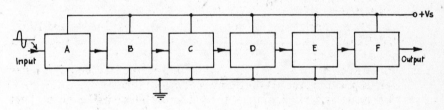

*Fig. 5.1*

(1) There is no output from Block F when a signal is applied to the input of Block A. The first test would be to check for output at:
    (a)    Block E
    (b)    Block C
    (c)    Block B
    (d)    Block A

(2) No output is obtained from Blocks F and D but an output is obtained from Block C. The fault is most likely confined to:
    (a)    Block C
    (b)    Block E
    (c)    Block F
    (d)    Block D

56    FAULT LOCATION IN ELECTRONIC EQUIPMENT

(3) There is no output from Blocks E or F but an output is obtained from Block C. The fault is most likely confined to:
   (a) Blocks D or E
   (b) Blocks E or F
   (c) Block E
   (d) Block B

(4) The signal is present at the input to Block A but there is no output from any of the blocks. Assuming that the d.c. supplies to each block are correct, the single fault probably lies in:
   (a) Block A
   (b) Block F
   (c) Block D
   (d) Blocks B–F

(5) There is no output from Block F when a signal is applied to the input of Block A. Amplifying units A–F are changed in turn but there is still no output from Block F. Assuming that a single fault is known to exist, where does the fault most probably lie?

Questions 6–10 refer to the audio amplifier block diagram of Fig. 5.2.

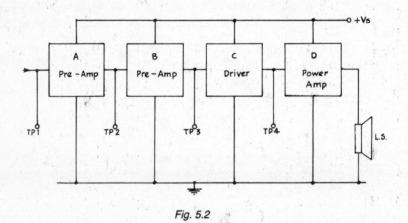

Fig. 5.2

(6) In order to develop the same standard output power in the loudspeaker, the largest signal level would be applied to:
   (a) TP1
   (b) TP2
   (c) TP3
   (d) TP4

# FAULT LOCATION EXERCISES

(7) A suitable frequency test tone for injection into the test points would be:
   (a) 2Hz
   (b) 2kHz
   (c) 200kHz
   (d) 2MHz

(8) When a test tone is injected into TP3 normal output is obtained, but distorted output is obtained when injecting at TP2 and TP1. Assuming that a single fault is present, the fault most likely lies in:
   (a) Block A
   (b) Block C
   (c) Block B
   (d) D.C. supply ($V_s$)

(9) Normal output is obtained when a test tone is injected into TP4 but there is no output when injection is made at TP3. The next step in the fault-finding procedure would be to:
   (a) Inject into TP2
   (b) Switch-off and measure the current drawn from the d.c. supply
   (c) Measure $V_s$
   (d) Switch-off and measure the d.c. conditions on the driver stage.

(10) A single fault is known to exist and during test signal injection the following signal amplitudes were required to raise the same standard output level in the loudspeaker:

   TP4–400 mV; TP3–20 mV; TP2–25 mV; TP1–1 mV

   The fault most likely lies in:
   (a) Block A
   (b) Block B
   (c) Block C
   (d) Block D

Questions 11–15 refer to the pulse generator system given in Fig. 5.3.

(11) The most suitable aid to isolate a fault to a single stage would be:
   (a) A c.r.o.
   (b) A digital multimeter
   (c) A logic probe
   (d) A logic analyser

# 58 FAULT LOCATION IN ELECTRONIC EQUIPMENT

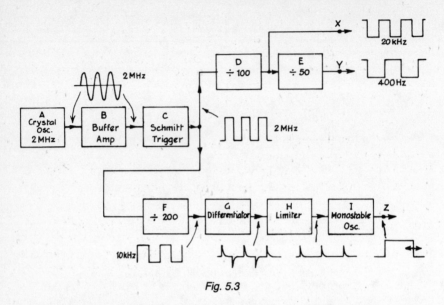

*Fig. 5.3*

(12) There is no output from points X, Y and Z. Assuming that a single fault is present, the fault is most likely confined to:
 (a) Block C
 (b) Blocks A, B or C
 (c) Block A
 (d) Blocks D or F

(13) Normal output is obtained at points X and Y but there is no output from point Z. The single fault present most probably lies in:
 (a) Block C
 (b) Block A
 (c) Block H
 (d) Blocks F–I

(14) Correct output is obtained at point X but there is no output at point Y or from Block F. These symptoms indicate that most likely:
 (a) Block C is faulty
 (b) Block E is faulty
 (c) Blocks D and E are faulty
 (d) Blocks E and F are faulty

(15) The output is correct at point Y but there is no output from point X. The fault present is probably due to:
 (a) A defective Block D
 (b) A faulty Block E
 (c) A s/c to the common line at the output of Block D
 (d) An o/c between Block D output and point X

# FAULT LOCATION EXERCISES

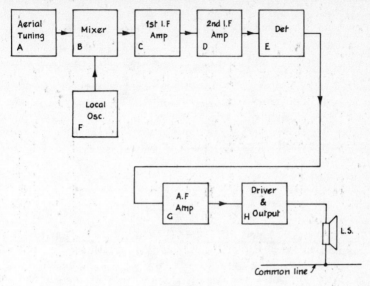

*Fig. 5.4*

Questions 16–20 refer to the a.m. transistor radio receiver block schematic of Fig. 5.4.

(16) The most suitable method for quickly isolating a fault to a particular stage is:
    (a)   Signal injection technique
    (b)   Stage-by-stage resistance checks
    (c)   D.C. voltage measurements on each stage
    (d)   Transistor substitution

(17) A suitable test frequency signal for injection into Block D would be:
    (a)   Unmodulated 1MHz
    (b)   Modulated 400Hz
    (c)   Modulated 470kHz
    (d)   Unmodulated 1MHz

(18) The following sensitivity readings were taken when a single fault was present for a standard power output level of 50mW:

| Injection point | Sensitivity |
| --- | --- |
| A.F. to input of Block G | 10mV |
| Mod. i.f. to output of Block D | 750mV |
| Mod. i.f. to input of Block D | 300mV |
| Mod. i.f. to output of Block C | 1V |
| Mod. i.f. to input of Block C | 80mV |
| Mod. i.f. to input of Block B | 10mV |

The conclusion to be made is that:
(a) Block B is faulty
(b) Block C is faulty
(c) Block D is faulty
(d) Block E is faulty

(19) A single fault is present. When modulated i.f. is applied to the input of Block B, normal output is obtained, but there is no output when modulated 200kHz (LW) and modulated 1MHz (MW) are applied at the input to Block B. The fault probably lies in:
(a) Block A
(b) Block F
(c) Block B
(d) Block E

(20) The receiver is tuned to 170kHz (LW) and a test signal of 170kHz is applied to the input of the mixer block. If Block F is functioning normally (assuming an i.f. of 470kHz), a c.r.o. connected at the local oscillator output would display a sinewave of frequency about:
(a) 170kHz
(b) 300kHz
(c) 470kHz
(d) 640kHz

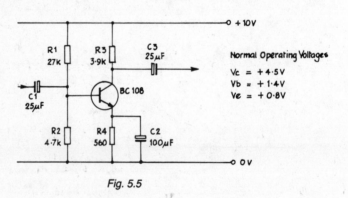

Fig. 5.5

Questions 21–25 refer to the single stage transistor amplifier of Fig. 5.5.

(21) Signal injection testing reveals that there is no output. Results of d.c. voltage measurements were:

$V_c = +10V$; $V_b = +1.4V$; $V_e = +1.0V$

Deduce the faulty component.

# FAULT LOCATION EXERCISES

(22) Signal injection tests reveal low gain. Results of d.c. voltage measurements were:

$V_c = +4.5V$; $V_b = +1.4V$; $V_e = +0.8V$

Deduce the most likely faulty component.

(23) Signal injection tests reveal no output. Results of d.c. voltage measurements were:

$V_c = +10V$; $V_b = 0V$; $V_e = 0V$

Deduce the faulty component.

(24) Signal injection tests reveal no output. Results of d.c. voltage measurements were:

$V_c = +10V$; $V_b = +0.2V$; $V_e = +0.2V$

Deduce the faulty component.

(25) Signal injection tests reveal severe distortion at output. Results of d.c. voltage measurements were:

$V_c = +1.4V$; $V_b = +2.1V$; $V_e = +1.4V$

Deduce the faulty component.

(26) The following fault conditions refer to the stabilised power supply of Fig. 5.6 which shows the normal on-load voltages at the various test points.

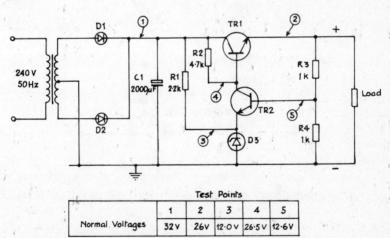

| Normal Voltages | 1 | 2 | 3 | 4 | 5 |
|---|---|---|---|---|---|
| Test Points | 32V | 26V | 12·0V | 26·5V | 12·6V |

All voltages measured with respect to the common earth line

Fig. 5.6

## 62  FAULT LOCATION IN ELECTRONIC EQUIPMENT

For each fault, deduce with reasons the single faulty component.

*Fault 1*

| Test Point 1 | 32V |
|---|---|
| Point 2 | 31·5V |
| Point 3 | 12V |
| Point 4 | 32V |
| Point 5 | 0V |

*Fault 2*

| Test Point 1 | 32V |
|---|---|
| Point 2 | 1·5V |
| Point 3 | 0V |
| Point 4 | 2V |
| Point 5 | 0·7V |

*Fault 3*

| Test Point 1 | 32V |
|---|---|
| Point 2 | 0V |
| Point 3 | 12V |
| Point 4 | 0V |
| Point 5 | 0V |

*Fault 4*

| Test Point 1 | 32V |
|---|---|
| Point 2 | 29V |
| Point 3 | 32V |
| Point 4 | 29·5V |
| Point 5 | 15V |

*Fault 5*

| Test Point 1 | 32V |
|---|---|
| Point 2 | 12·7V |
| Point 3 | 12V |
| Point 4 | 13·2V |
| Point 5 | 12·6V |

(27) The following fault conditions refer to the Monostable Multivibrator of Fig. 5.7 which shows the d.c. voltages at the test points **without** trigger applied and measured with an instrument of 1MΩ input impedance.

For each fault, deduce with reasons the most likely faulty component(s). In all cases the circuit does not operate and no substantial pulses are obtained at the output. The d.c. supply rails are assumed to be present.

# FAULT LOCATION EXERCISES

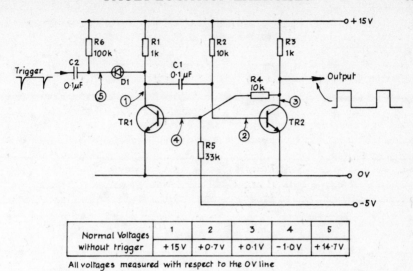

| Normal Voltages without trigger | 1 | 2 | 3 | 4 | 5 |
|---|---|---|---|---|---|
| | +15 V | +0·7 V | +0·1 V | −1·0 V | +14·7 V |

All voltages measured with respect to the 0V line

*Fig. 5.7*

*Fault 1*

    Test Point 1    + 15V
    Point 2    + 0·7V
    Point 3    + 0·1V
    Point 4    − 4·7V
    Point 5    + 14·7V

*Fault 2*

    Test Point 1    + 15V
    Point 2    + 0·7V
    Point 3    + 0·1V
    Point 4    − 1·0V
    Point 5    + 14·7V

*Fault 3*

    Test Point 1    + 0·18V
    Point 2    0V
    Point 3    + 13·7V
    Point 4    + 0·7V
    Point 5    + 12·3V

*Fault 4*

    Test Point 1    0V
    Point 2    + 0·7V
    Point 3    + 0·1V
    Point 4    − 1·0V
    Point 5    + 12·3V

# FAULT LOCATION IN ELECTRONIC EQUIPMENT

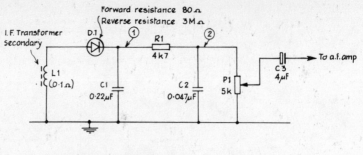

Normal Test Point Resistances

|  | 1 | 2 |  |
|---|---|---|---|
| Ohmmeter leads one way round | 9·7 kΩ | 5 kΩ | Measured between Test Point and chassis line |
| Ohmmeter leads reversed | 79 Ω | 2·4 kΩ |  |

*Fig. 5.8*

(28) The following faults refer to the radio receiver demodulator of Fig. 5.8 which also gives a table of normal test point resistances.

Signal injection tests have narrowed down the area of the fault to the circuit shown and in each case there is no output from the receiver. Deduce the faulty component. The second resistance reading given at each test point shows the effect of reversing the ohmmeter lead connections.

*Fault 1*

| Test Point 1 | 3 MΩ |
|  | 80 Ω |
| Test Point 2 | 5 kΩ |
|  | 5 kΩ |

*Fault 2*

| Test Point 1 | 4·7 kΩ |
|  | 79 Ω |
| Test Point 2 | 0 Ω |
|  | 0 Ω |

*Fault 3*

| Test Point 1 | 3 MΩ |
|  | 81 Ω |
| Test Point 2 | 3 Ω |
|  | 4·8 kΩ |

## FAULT LOCATION EXERCISES

*Fault 4*

| | | |
|---|---|---|
| Test Point 1 | | 9.7kΩ |
| | | 79Ω |
| Test Point 2 | | 5kΩ |
| | | 2.4kΩ |

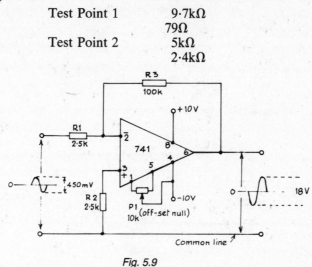

*Fig. 5.9*

(29) Refer to Fig. 5.9 which shows the circuit of an op-amp inverting amplifier. With the input s/c to the common line, P1 is adjusted for zero output voltage on pin 6. When signal is applied with a peak-to-peak amplitude of 450mV an output of 18V peak-to-peak is obtained.

State the probable effect on the output when the following faults are present:
(a) R1 o/c
(b) R3 o/c
(c) R2 o/c

(30) Refer to Fig. 5.10 which shows the circuit of an op-amp summing amplifier. With both inputs s/c to the common line, P1 is adjusted for

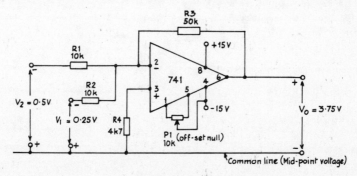

*Fig. 5.10*

zero output voltage. When inputs V1 and V2 are applied with amplitudes and polarities as shown, an output of 3·75V with polarity as indicated is obtained.

State the probable effect on the output when the following faults are present:
(a) R1 o/c
(b) R2 o/c
(c) R3 o/c

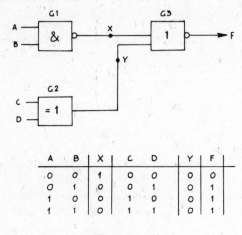

Fig. 5.11

(31) Refer to Fig. 5.11 which shows a combinational logic circuit. The truth table gives the logic state of the gate outputs when a single fault is present. Deduce the location and nature of the fault.

(32) Refer to Fig. 5.12 which shows a simple l.e.d. control circuit.

The l.e.d. fails to emit light for all combinations of inputs A and B and the truth table lists the voltage levels under the fault conditions. Deduce the most likely cause(s) of the fault.

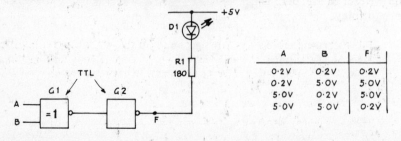

Fig. 5.12

# FAULT LOCATION EXERCISES

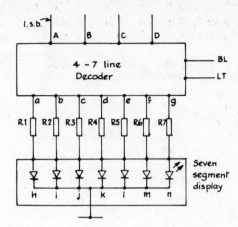

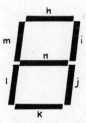

Fig. 5.13

(33) Refer to Fig. 5.13 which shows a 4–7 line decoder and seven-segment display.

The following table gives the logic state for two separate combinations of the decoder outputs a–g, where H = + 5V and L = 0V. In both cases numeral 5 is displayed. Deduce the cause(s) of the fault.

| a | b | c | d | e | f | g |
|---|---|---|---|---|---|---|
| H | L | H | H | L | H | H |
| H | L | H | H | H | H | H |

# 68  FAULT LOCATION IN ELECTRONIC EQUIPMENT

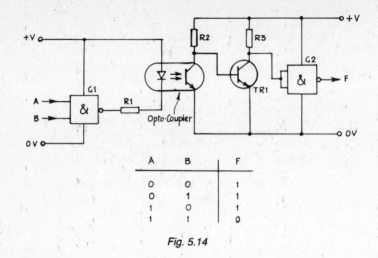

| A | B | F |
|---|---|---|
| 0 | 0 | 1 |
| 0 | 1 | 1 |
| 1 | 0 | 1 |
| 1 | 1 | 0 |

Fig. 5.14

(34) Refer to Fig. 5.14 which shows a logic circuit interface and lists the normal logic state at the output F for the combinations of inputs A and B.

Deduce the effect on the logic state at F when the following faults are present:
(a)   R3 o/c
(b)   R2 o/c
(c)   TR1 base–emitter s/c
(d)   R1 o/c

(35) Refer to Fig. 5.15 which shows the circuit of a unijunction transistor oscillator and gives the d.c. readings at the various test points when the circuit is functioning normally.

In each of the following faults the circuit fails to oscillate and the d.c. voltage readings obtained were:

*Fault 1*

        Test Point 1        0V
                    2        9·6V
                    3        0V
                    4        0V

*Fault 2*

        Test Point 1        0·8V
                    2        0V
                    3        0V
                    4        0·8V

# FAULT LOCATION EXERCISES

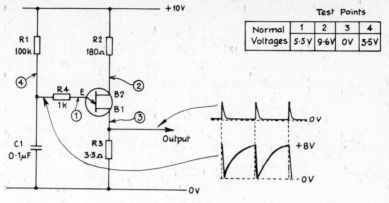

Fig. 5.15

Fault 3

| Test Point | 1 | 0V |
|---|---|---|
| | 2 | 9·6V |
| | 3 | 0V |
| | 4 | 9V |

Fault 4

| Test Point | 1 | 9V |
|---|---|---|
| | 2 | 9·9V |
| | 3 | 9·6V |
| | 4 | 9V |

(36) Refer to Fig. 5.16 which shows a positional servo system.

State the probable effect on the operation of the system with the following faults present:
(a) R1 o/c
(b) R2 o/c
(c) R3 o/c
(d) P3 slider set too high
(e) P3 slider set too low

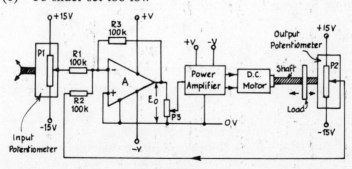

Fig. 5.16

# ANSWERS TO EXERCISES

1. (b) Use 'half-split' method.
2. (d)
3. (a)
4. (a)
5. In the supply line $V_s$. Either the supply line voltage is absent or there is a disconnection or s/c in the common feed to all of the blocks.
6. (d)
7. (b) Mid-audio range.
8. (c) Injection at TP3 reveals that blocks C and D are normal. Distortion is first noticed at TP2, thus the distortion is occurring somewhere between TP2 and TP3, most probably in block B. Injection at TP1 will increase the distortion due to the increased signal level arriving at the input to Block B.
9. (d) The limit of dynamic testing has been reached.
10. (b) The injection signal levels required show that there is no gain in block B. The actual gain is only 20/25 = 0·8, which represents a signal loss.
11. (a)
12. (b) Fault most likely confined to the common blocks prior to divergence.
13. (d) Since outputs at X and Y are normal, output from block C must be normal.
14. (d)
15. (d)
16. (a)
17. (c) Modulated i.f.
18. (c) The sensitivity readings reveal low gain in block D (750/300 = 2·5).
19. (b) Injection of modulated i.f. at input to block B reveals it is working as an amplifier. The signal injection tests at 200kHz and 1MHz indicate that there is no frequency-changing action, thus the fault probably lies in the local oscillator.
20. (d) Local oscillator tuned above the incoming signal by 470kHz.

# FAULT LOCATION EXERCISES – ANSWERS

21  R4 o/c
22  C2 o/c
23  R1 o/c         Explanations of the faults in exercises 21–25 are
24  Base–emitter s/c   given in Chapter 3.
25  R2 o/c
26  *Fault 1*
There is an output voltage but no voltage at the junction of R3/R4. This points to either a s/c across R4 or an o/c R2. A short-circuited R4 is not a practical possibility (unless the printed circuit is shorted), thus R3 is open-circuited.
*Fault 2*
The clue to this fault is the 0V at test point 3, indicating a short-circuit across D3. The most likely fault is D3 s/c.
*Fault 3*
There is no voltage at TR2 collector but a voltage is present at test point 1, thus R2 is open-circuit. Note that with zero voltage at TR1 base there will be zero voltage at TR1 emitter and hence zero voltage at TR2 base.
*Fault 4*
The output voltage is high indicating that stabilising action is not taking place. The voltage across D3 is high thus pointing to an o/c D3.
*Fault 5*
Since the voltage at the junction of R3/R4 is almost the same as that at test point 2, there is no potential divider action taking place. There is little voltage drop across R3 indicating little current in it, thus pointing to an o/c R4.

27  *Fault 1*
The only voltage reading which is different is that at test point 4. The voltage at this point is the result of a potential divider action of R4 and R5 between the voltage at test point 3 and the − 5V supply. Test point 3 has the correct voltage and assuming the − 5V supply to be correct, R4 is most likely o/c.
*Fault 2*
All the d.c. voltages (without trigger) are correct thus indicating an a.c. type fault or a loss of trigger. This could be due to an o/c C1, o/c C2 or o/c D1.
*Fault 3*
Under normal operating conditions TR2 is 'on' and TR1 is 'off' so that the voltage at test point 3 is low and that at test point 1 is high. The voltages given at test points 2 and 3 indicate that TR2 is 'off'. The high voltage at TR2 collector will cause TR1 to come 'on' and for its collector voltage to be low. The clue is the presence of zero voltage at test point 2. This could be caused by either an o/c R2 or TR2 base–emitter s/c.
*Fault 4*
The clue to this fault is zero voltage at test point 1, all other voltages being approximately correct. Zero voltage at test point 1 indicates

either a short-circuit to the 0V rail or an open-circuit to the + 15V rail. A short-circuit between TR1 collector and emitter usually results from collector, base and emitter all being shorted together. This cannot be the case, as the base of TR1 is at − 1·0V. Thus the most likely cause is R1 o/c.

28   Under normal conditions with the ohmmeter connected between test point 1 and chassis, the ohmmeter will read the forward resistance of D1 in parallel with the series combination of R1 and P1; with the meter leads reversed it will read the reverse resistance of D1 in parallel with R1 and P1 in series.

When connected between test point 2 and chassis the ohmmeter will read the resistance of P1 in parallel with the series combination of R1 and the forward or reverse resistance of D1, depending upon the polarity of the ohmmeter leads.

*Fault 1*
The resistance readings at test point 2 are that of P1 only thus indicating an open-circuit in either R1, D1 or L1. However the readings at test point 1 give the forward or reverse resistances of D1, revealing a resistance path through D1 and L1. Thus R1 must be open-circuit.

*Fault 2*
The clue to this fault is zero ohms at test point 2 pointing to C2 being s/c. This is confirmed by the lower resistance obtained at test point 1 when the diode is reverse biased.

*Fault 3*
The readings obtained point to P1 being o/c. Since the symptoms are no output, the fault cannot be P1 open-circuit at the lower end of its track as this would result in non-controllable full volume. Thus P1 is o/c at the top end of its track.

*Fault 4*
The resistance readings obtained are normal, thus possible faults causing no output are C3 o/c or P1 slider o/c.

29   (a)  With R1 o/c there will be no output signal as there will be no signal input to the inverting terminal.
      (b)  Normally the gain of the amplifier will be equal to $R3/R1 = 100/2·5 = 40$ (closed loop). With R3 o/c the gain of the op-amp will be very large (about 200,000 on open-loop) resulting in severe clipping of the output as it reaches the limit of the d.c. supply voltage of ± 10V.
      (c)  With R2 o/c the non-inverting input will be floating. This will upset the balance of the differential input amplifier, resulting in no output and a large d.c. off-set voltage at the output.

30   (a)  Note that the voltage gain to the two inputs is $R3/R1$ and $R3/R2 = 5$. With R1 o/c the output voltage will be $5 \times 0·25 = 1·25V$ with the output positive to the common line.
      (b)  With R2 o/c the output voltage will be $5 \times 0·5 = 2·5V$ with the output positive to the common line.

# FAULT LOCATION EXERCISES – ANSWERS

(c) With R3 o/c the voltage gain will be very large (open-loop) resulting in the output voltage swinging in the positive direction and limiting at practically the positive supply line of + 15V.

31 From the truth table it will be seen that the output at Y of the EX-OR gate G2 incorrectly assumes the logic 0 state for all input combinations. The output of the NAND gate at X is correct for inputs A and B. Also G3 output is correct for inputs X and Y. Thus G2 is faulty with its output permanently stuck at logic 0.

32 The truth table shows the correct voltage levels at point F for the input combinations, indicating that the gates G1 and G2 are functioning correctly. The l.e.d. should emit light when the output at F is **low** at 0·2V thus forward-biasing the diode. Thus possible faults are R1 o/c or D1 o/c (assuming the positive supply rail voltage is present).

33 The first combination of decoder output states is that required to produce numeral 5 on the display. The second combination is that required to produce numeral 6. Since numeral 5 is displayed, segment 1 is failing to illuminate. Thus possible faults are R5 o/c or diode 1 o/c.

34 Under normal operating conditions with inputs A and B both **high** at logic 1, the output of G1 is **low** at logic 0 and the l.e.d. of the opto-coupler is forward-biased and emitting light. The light falls on the photo transistor of the opto-coupler causing it to conduct and for its collector voltage to go **low**. This results in TR1 turning 'off' and for its collector voltage to go **high**. After inversion by G2, output F assumes the **low** state.

(a) With R3 o/c the output at F will assume the **high** state for all input combinations.

(b) With R2 o/c, TR1 will be permanently 'off' causing the output at F to be **low** regardless of the logic state of the inputs.

(c) With TR1 base–emitter short, TR1 will be permanently 'off' causing the output at F to be **low** regardless of the logic state of the inputs.

(d) With R1 o/c the l.e.d. will fail to conduct for the input condition $A = B = 1$, in which case the output at F will be **high** for all input combinations.

35 Under normal operating conditions at switch-on, C1 charges exponentially via R1 towards the supply rail voltage. Provided the voltage across C1 is less than 0·4–0·8 of the voltage between B1 and B2, the u.j.t. remains 'off'. When the rising voltage across C1 exceeds 0·8 of the voltage between B1 and B2 in this case, the emitter becomes conducting and C1 is discharged rapidly as current flows in the emitter-to-B1 junction. This results in a small positive pulse in the output.

*Fault 1*
The clue to this fault is the zero voltage at test point 4 and could be caused by either an open-circuit R1 or a short-circuit C1, both faults providing identical voltage readings.

*Fault 2*
The 0V at point 2 indicates that R2 is most likely o/c. Removing the

voltage from B2 causes the emitter to conduct at the low voltage of 0·8V and no oscillatory action takes place.
*Fault 3*
The high voltage at point 4 and the low voltage at point 1 indicate that R4 is o/c. If point 1 were s/c the voltage at point 4 would be low due to the small value of R4 compared with R1. The voltage at point 4 is high as C1 fully charges to the supply rail; due to the effect of voltmeter loading it is less with the voltmeter connected.
*Fault 4*
The clue to this fault is the high voltage at point 3 and indicates that R3 is open-circuit.

36  Normal operation is as follows:
Suppose that initially P1 and P2 slider outputs are at zero voltage and the input potentiometer is set to $-2V$. As P1 is at $-2V$ and P2 at 0V, the sum is $-2V$, causing the summing amplifier A to produce an inverted d.c. output voltage ($E_0$) proportional to the sum. This voltage is applied via P3 to the power amplifier which drives the motor. The torque developed by the motor turns the shaft and moves the load towards the desired position. This causes P2 slider to move towards the positive supply rail which reduces the sum input to A. There is less error voltage ($E_0$) developed thus less power is fed to the motor, but it will continue to accelerate, moving the load and increasing the positive potential supplied from P2 slider. When the voltage from P2 is equal to $+2V$, the sum input to A will be zero. In consequence there will be zero error voltage, no power will be supplied to the motor and no torque will be developed. The load will be in the desired position set by P1.

(a) As soon as R1 goes o/c, amplifier A produces an error voltage out (provided P2 output is not zero) causing the load to move to a position such that the output from P2 is zero. The motor will then be stopped and the input potentiometer will have no effect on the system.

(b) With R2 o/c there will be no feedback to the summing amplifier. In consequence, as P1 slider voltage is altered in either the positive or negative direction, the motor will drive the load to one of its extreme end positions. Under this fault condition damage to the motor or power amplifier may result unless power drive to the motor is automatically shut down.

(c) With R3 o/c the gain of the summing amplifier will be large (on open-loop) causing a large error voltage output when P1 is adjusted. This will result in the load oscillating about its desired position.

(d) A similar effect to (c) occurs if the setting of P3 is too high. With excessive drive to the motor the load will overshoot the desired position when P1 is adjusted and oscillate about that position. P3 is adjusted normally to give stable operation.

(e) If P3 is set too low there may be insufficient drive to turn the motor and move the load as P1 setting is altered.